STUDY GUIDE FOR

ENGINEERING ECONOMIC ANALYSIS

FOURTEENTH EDITION

Donald G. Newnan

Ted G. Eschenbach

Jerome P. Lavelle

Neal A. Lewis

PREPARED BY

Ed Wheeler

Neal A. Lewis

Ted G. Eschenbach

New York   Oxford

Oxford University Press

D0907673

Oxford University Press is a department of the University of Oxford. It furthers the University's objective of excellence in research, scholarship, and education by publishing worldwide. Oxford is a registered trade mark of Oxford University Press in the UK and certain other countries.

Published in the United States of America by Oxford University Press
198 Madison Avenue, New York, NY 10016, United States of America.

ISBN 978-0-19-093200-8

9 8 7 6 5 4 3 2 1
Printed by Marquis, Canada

Contents

Foreword

Engineering economics is learned by solving problems, not by reading how to solve problems. Students sometimes need additional problems to solve, along with their solutions, beyond their homework assignments. As a textbook evolves, materials supporting the text need to evolve with it. Old tests, old course notes, and old books do not always meet the need. This study guide was created to fill that need, and it has evolved over the years with our textbook. We are happy to present this new study guide which was edited and updated to support the fourteenth edition of *Engineering Economic Analysis*.

This new edition of the study guide contains over 400 problems, with 30% being new or revised. Spreadsheet solutions are used for many of the new problems. Depreciation and tax problems have been updated to reflect the new tax laws, including revised depreciation schedules and tax rates.

I would like to thank Dr. Ed Wheeler of the University of Tennessee at Martin for his many years of work to bring you this study guide. I would also like to thank Dan Sayre of Oxford University Press for his support in continuing the evolution of this guide.

For the *Engineering Economic Analysis* author team,

Neal Lewis

Formulas

Compound Amount: To find F, given P

$(F/P, i, n)$ $F = P(1+i)^n$

Present Worth: To find P, given F

$(P/F, i, n)$ $P = F(1+i)^{-n}$

Series Compound Amount: To find F, given A

$(F/A, i, n)$ $F = A\left[\dfrac{(1+i)^n - 1}{i}\right]$

Sinking Fund: To find A, given F

$(A/F, i, n)$ $A = F\left[\dfrac{i}{(1+i)^n - 1}\right]$

Capital Recovery: To find A, given P

$(A/P, i, n)$ $A = P\left[\dfrac{i(1+i)^n}{(1+i)^n - 1}\right]$

Series Present Worth: To find P, given A

$(P/A, i, n)$ $P = A\left[\dfrac{(1+i)^n - 1}{i(1+i)^n}\right]$

Arithmetic Gradient Uniform Series: To find A, given G

$(A/G, i, n)$ $A = G\left[\dfrac{(1+i)^n - in - 1}{i(1+i)^n - i}\right]$ or $A = G\left[\dfrac{1}{i} - \dfrac{n}{(1+i)^n - 1}\right]$

Arithmetic Gradient Present Worth: To find P, given G

$(P/G, i, n)$ $P = G\left[\dfrac{(1+i)^n - in - 1}{i^2(1+i)^n}\right]$

Geometric Gradient: To find P, given A_1, g

$(P/G, g, i, n)$ $P = A_1\left[n(1+i)^{-1}\right]$ $P = A_1\left[\dfrac{1 - (1+g)^n(1+i)^{-n}}{i - g}\right]$

                       when $i = g$ when $i \neq g$

Compound Interest

i = Interest rate per interest period

n = Number of interest periods

P = A present sum of money

F = A future sum of money

A = An end-of-period cash receipt or disbursement in a uniform series continuing for n periods

G = Uniform period-by-period increase or decrease in cash receipts or disbursements

g = Uniform rate of cash flow increase or decrease from period to period; the geometric gradient

r = Nominal interest rate per interest period

m = Number of compounding subperiods per period

Effective Interest Rates

For noncontinuous compounding: i_{eff} or $i_a = \left(1 + \dfrac{r}{m}\right)^m - 1$

where r = nominal interest rate per year
m = number of compounding periods in a year

OR

$$i_{\text{eff}} \text{ or } i_a = (1 + i)^m - 1$$

where i = effective interest rate per period
m = number of compounding periods in a year

For continuous compounding: i_{eff} or $i_a = (e^r) - 1$

where r = nominal interest rate per year

Values of Interest Factors When n *Equals Infinity*

Single Payment:

$(F/P, i, \infty) = \infty$

$(P/F, i, \infty) = 0$

Uniform Payment Series:

$(A/F, i, \infty) = 0$ $(F/A, i, \infty) = \infty$

$(A/P, i, \infty) = i$ $(P/A, i, \infty) = 1$

Arithmetic Gradient Series:

$(A/G, i, \infty) = 1/i$

$(P/G, i, \infty) = 1/i^2$

Chapter 1

Making Economic Decisions

1-1

Many engineers earn high salaries for creating profits for their employers and then, at retirement time, find themselves insufficiently prepared financially. This may be because in college courses there is little or no discussion on using engineering economics for the direct personal benefit of the engineer. Among the goals of every engineer should be assuring that adequate funds will be available for anticipated personal needs at retirement.

A realistic goal of retiring at age 65 with a personal net worth in excess of $2 million can be accomplished by several methods. An independent study ranked the probability of success of the following methods of personal wealth accumulation. Discuss and decide the ranking order of the following five methods.

a. Purchase as many lottery tickets as possible with money saved from salary.
b. Place money saved from salary in a bank savings account.
c. Place all money saved from a salary in a money market account.
d. Invest saved money into rental properties and spend evenings, weekends, and vacations repairing and managing.
e. Invest all money saved into stock market securities, and study investments 10 to 15 hours per week.

Solution

Independent studies can be misleading. If a recent winner of a $2 million lottery drawing were asked to rank wealth accumulation methods, item a would head his or her list. A workaholic with handyman talent might put item d as his Number 1 choice. Similarly, many people have become millionaires by investing in real estate and in other ways not listed here. The important thing is to learn about the many investment vehicles available and then choose the one or the several most suitable for you.

1-2

The following letter to Joseph Priestley, the English chemist, was written by his friend Benjamin Franklin. Priestley had been invited to become the librarian for the Earl of Shelburne and had asked for Franklin's advice. What engineering economy principle does Franklin suggest Priestley use to aid in making his decision?

London, September 19, 1772

Dear Sir:

In the affair of so much importance to you wherein you ask my advice, I cannot, for want of sufficient premises, advise you what to determine, but if you please I will tell you how. When these difficult cases occur, they are difficult chiefly because while we have them under consideration, all the reasons Pro and Con are not present to the mind at the same time; but sometimes one set present themselves, and at other times another, the first being out of sight. Hence the various purposes or inclination that alternately prevail, and the uncertainty that perplexes us.

To get over this, my way is to divide a half a sheet of paper by a line into two columns; writing over the one PRO and over the other CON. Then during three or four days' consideration I put down under the different heads short hints of the different motives that at different times occur to me, for or against the measure. When I have thus got them all together in one view, I endeavour to estimate their respective weights; and where I find two (one on each side) that seem equal, I strike them both out. If I find a reason Pro equal to some two reasons Con, I strike out the three. If I judge some two reasons Con equal to three reasons Pro, I strike out the five; and thus proceeding I find at length where the balance lies; and if after a day or two of further consideration, nothing new that is of importance occurs on either side, I come to a determination accordingly. And though the weight of the reasons cannot be taken with the precision of algebraic quantities, yet when each is thus considered separately and comparatively and the whole lies before me, I think I can judge better, and am less likely to make a rash step; and in fact I have found great advantage from this kind of equation in what may be called moral or prudential algebra.

Wishing sincerely that you may determine for the best, I am ever, my dear friend, yours most affectionately. . .

s/Ben Franklin

Solution

Decisions should be based on the differences between the alternatives. Here the alternatives are taking the job (Pro) and not taking the job (Con).

1-3

Assume that you are employed as an engineer for Wreckall Engineering Inc., a firm specializing in the demolition of high-rise buildings. The firm has won a bid to tear down a 30-story building in a heavily developed downtown area. The crane the company owns reaches only to 29 stories. Your boss asks you to perform an economic analysis to determine the feasibility of buying a new crane to complete the job. How would you handle the analysis?

Solution

The important point of this exercise is to realize that your boss may not have recognized what the true problem is in this case. To buy a new crane is only <u>one</u> alternative and quite likely not the best alternative. Others include:

   extension on current crane
   ramp for current crane
   renting a crane to remove the top story
   explosive demolition

If this is a fixed-output project (for example, there is a fixed fee for demolishing building) we want to minimize costs. Weigh alternatives using economic criteria to choose the best alternative.

1-4

By wisely saving and investing, Helen finds she has accumulated $400,000 in savings while her salaried position is providing her with $40,000 per year, including benefits, after income taxes and other deductions.

Helen's salaried position is demanding and allows her little free time, but the desire to pursue other interests has become very strong. What would be your advice to her, if you were asked?

Solution

First, Helen should decide what annual income she needs to provide herself with the things she wants. Depending on her age, she might be able to live on the interest income (maybe $10\% \times \$400,000 = \$40,000$), or a combination of interest and principal. The important thing for Helen to realize is that it may be possible for her to have a more fulfilling lifestyle if she is fully aware of the time value of money. Many people keep large sums of money in bank checking accounts (drawing no interest) because they can write "free" checks.

1-5

Charles belongs to a square dance club that meets twice each month and has quarterly dues of $9 per person. The club moved its meeting place to a more expensive location. To offset the increased cost, members agreed to pay 50 cents apiece each time they attend the meeting. Later the treasurer suggests that the quarterly dues be increased to $12 per person as an alternative to the meeting charge. Discuss the consequences of the proposal. Do you think all the club members will agree to the proposal?

Solution

The members who attend regularly would pay the same amount with the new dues as with the older method: that is, $9 plus 50 cents per meeting. Many would like the added advantage of covering their quarterly expenses in one check. The members who attend infrequently would pay more under the new method and might oppose the action.

Since the people who attend infrequently are in the minority in this club, the proposal was approved when the members voted.

1-6

A PhD student accepted a full-time teaching job in February, with the job starting in June. Two weeks before the job was to start, this person received another job offer at a larger university, paying 10% more. Should they accept this new offer or turn it down?

Solution

This occurs more often than you might think. There are some serious concerns that need to be considered.

- When the student accepted the offer in February, they made a verbal contract (and perhaps a written one). If the second offer is accepted, then the student has broken a contract. The student's integrity has been compromised, and legal action by the first university could result.
- The student should ask how they might feel if they were in the position of the first university.
- If the student accepts the second offer, they have poisoned their relationship with the first university for years to come; possibly for their entire career. The student's reputation could suffer.

1-7

A coal-fired power plant produces electricity for the region. They have been told they need to reduce their emissions of mercury due to the toxic nature of their smoke emissions. The plant says that they cannot afford to reduce mercury, and that people will be laid off if they need to reduce production in order to meet emission standards. What should they do?

Solution

> This is an ongoing debate. There is clear evidence that people, especially children, living downwind of such power plants are seriously harmed by the mercury emissions from coal-fired power plants. At the same time, the employment of people can be critical to the community near the plant also. These two points are often the key arguments in the debate. Many coal-fired power plants are being shut down or converted to natural gas, not for health reasons, but for economic reasons (natural gas is cheaper).

1-8

Car *A* initially costs $500 more than Car *B*, but it consumes 0.04 gallon/mile versus 0.05 gallon/mile for *B*. Both vehicles last 8 years, and *B*'s salvage value (the value when it is traded in after 8 years) is $100 smaller than *A*'s. Fuel costs $2.40 per gallon. Other things being equal, at how many miles of use per year (*X*) is *A* preferred vs. *B*?

Solution

$$-500 + 100 + (0.05 - 0.04)\,(2.40)\,(8)X = 0$$
$$-400 + 0.192X = 0$$
$$X = 400/0.192 = 2083 \text{ miles/year}$$

1-9

Sam decides to buy a cattle ranch and leave the big-city rat race. He locates an attractive 500-acre spread in Montana for $1000 per acre that includes a house, a barn, and other improvements. Sam's studies indicate that he can run 200 cow–calf pairs and be able to market 180 500-pound calves per year. Sam, being rather thorough in his investigation, determines that he will need to purchase an additional $95,000 worth of machinery. He expects that supplemental feeds, medications, and veterinary bills will be about $50 per cow per year. Property taxes are $4000 per year, and machinery upkeep and repairs are expected to run $3000 per year.

If interest is 10% and Sam wants a net salary of $10,000 per year, how much must each 500-pound calf sell for?

Solution

Land cost: $500 acres × $1000/acre = $500,000
Machinery: Lump sum = <u>95,000</u>
Total fixed cost $595,000
Assume land and machinery to have a <u>very</u> long life:
at 10% annual cost = (0.10)($595,000) = $59,500
Other annual costs:

      Feed, medications, vet bills: $50 × 200 = $10,000
      Property taxes   4,000
      Upkeep and repairs   3,000
      Salary   <u>10,000</u>
Total annual cost $86,500

Net sale price of each calf would have to be $86,500/180 = $480.56.
Note: If Sam were to invest his $595,000 in a suitable investment vehicle yielding 10% interest, his salary would be almost six times greater, and he could go fishing instead of punching cows.

1-10

A food processor is considering the development of a new product. Depending on the quality of raw material, he can expect different yields process-wise, and the quality of the final products will also vary considerably. The product development department has identified three alternatives, which it has produced on a pilot scale. The marketing department has used those samples for surveys to estimate potential sales and pricing strategies. The three alternatives, which would use existing equipment, but different process conditions and specifications, are summarized as follows. Indicate which alternative seems to be the best according to the estimated data, if the objective is to maximize total profit per year.

	Alternative		
	1	2	3
Pounds of raw material *A* per unit of product	0.05	0.07	0.075
Pounds of raw material *B* per unit of product	0.19	0.18	0.26
Pounds of raw material *C* per unit of product	0.14	0.12	0.17
Other processing costs ($/unit product)	$0.16	$0.24	$0.23
Expected wholesale price ($/unit product)	0.95	1.05	1.25
Projected volume of sales (units of product)	1,000,000	1,250,000	800,000

Cost of raw material *A* $3.45/lb
Cost of raw material *B* $1.07/lb
Cost of raw material *C* $1.88/lb

Solution

		Alternative		
		1	2	3
Cost of raw material A ($/unit product)	$0.05 \times 3.45 =$	0.1725	0.2415	0.2587
Cost of raw material B ($/unit product)	$0.19 \times 1.07 =$	0.2033	0.1926	0.2782
Cost of raw material C ($/unit product)	$0.14 \times 1.88 =$	0.2632	0.2256	0.3196
Other processing costs ($/unit product)		$0.16	$0.24	$0.23
Total cost ($/unit product)		0.799	0.8997	1.0865
Wholesale price ($/unit product)		0.95	1.05	1.25
Profit/unit		0.151	0.1503	0.1635
Projected sales (units of product)		1,000,000	1,250,000	800,000
Projected profits		151,000	187,875	130,800

Therefore, choose Alternative 2.

1-11
The total cost of a building (TC) is given by

$$TC = \left(200 + 80X + 2X^2\right)A$$

where X = number of floors and A = floor area (ft^2 per floor)
If the total number of square feet required is 10^6, what is the optimal (minimum cost) number of floors?

Solution

$$TC = \left(200 + 80X + 2X^2\right)\left(\frac{10^6}{X}\right)$$

$$\frac{dTC}{dx} = \left(10^6\right)\left(\frac{-200}{X^2} + 2\right) = 0$$

$$X^* = \sqrt{\frac{200}{2}} = \sqrt{100} = 10 \text{ floors}$$

1-12

Two locations are being considered for a new regional office. Factors to consider include the cost of living, utilities and taxes, and quality of life. Each has been rated, as shown:

Factor	Location A	Location B	Factor weight
Cost of living	3	5	25%
Utilities and taxes	4	4	40%
Quality of life	5	3	35%

Given the data shown, which location should be chosen?

Solution

Weighted score for Location A $= 0.25(3) + 0.40(4) + 0.35(5) = 4.10$

Weighted score for Location B $= 0.25(5) + 0.40(4) + 0.35(3) = 3.90$

Choose Location A because it has the higher score.

1-13

A new warehouse is being planned, and 3 locations are being compared. Factors being considered include local labor cost, taxes, and access to interstate highways. These are summarized in the table as follows:

Factor	Location 1	Location 2	Location 3	Factor weight
Labor cost	7	6	5	35%
Taxes	8	7	6	25%
Highway access	3	5	8	40%

Which location should be selected?

Solution

Weighted score for Location 1 $= 0.35(7) + 0.25(8) + 0.40(3) = 5.65$

Weighted score for Location 2 $= 0.35(6) + 0.25(7) + 0.40(5) = 5.85$

Weighted score for Location 3 $= 0.35(5) + 0.25(6) + 0.40(8) = 6.45$

Choose Location 3 because it has the higher score.

Chapter 2
Estimating Engineering Costs and Benefits

2-1

DC Brick has fixed cost of production of $2,562,500 per year and a unit production cost of $0.75. Marvel Brick has a fixed cost of production of $1,000,000 and a unit production cost of $2.00. At what number of units produced will the two companies have equal annual production costs?

Solution

At equal annual production cost $\quad 2,562,500 + 0.75X = 1,000,000 + 2.00$

$$1,562,500 = 1.25X$$
$$X = 1,250,000 \text{ units}$$

2-2

A household is considering installing solar panels, but they need to understand what they are paying for electricity. They use an average of 15 kilowatt hours of electricity per day, and an average month is 30 days. They pay $0.22 per kilowatt hour for the first 350 kilowatt hours used in a month, and $0.19 per kilowatt hour after that.

a. What is their average cost of electricity?

b. What is their marginal cost of electricity?

Solution

a. Average monthly usage = (15 kw-hr/day)(30 days) = 450 kw-hr

$$\text{Average cost} = \frac{(0.22)(350) + (0.19)(450 - 350)}{450} = \frac{77.00 + 19.00}{450}$$
$$= \$0.2133 \text{ per kilowatt hour}$$

b. Marginal cost = $0.19/kw-hr

2-3

Your Econ. Professor bought airline tickets to leave a conference early on Saturday morning. He later found out that he needed to present a paper later that Saturday, so he changed his tickets at a cost of $200. When he got to the conference, he discovered that he was scheduled twice: on Friday and on Saturday. He was given his choice of when to give the presentation. He said he would present on Saturday, because he paid an extra $200 in order to stay. Explain why you agree or disagree with his logic.

Solution

Whether he presents on Friday or Saturday does not change the fact that he paid the $200, which is a sunk cost. There may be other reasons to select one day or the other, but the sunk cost should not impact the decision.

2-4

LED lights are replacing incandescent lighting in many applications, including in industrial application. LED lights cost more (but prices continue to decrease), last longer, and consume less electricity. Identify fixed and variable costs that need to be considered when doing an economic analysis.

Solution

Fixed costs would include any fixtures that might need to be changed.

Variable costs include the higher cost of new LED bulbs, decreased labor cost to change burned out bulbs, and decreased electricity costs.

A nuclear power plant switched to LED bulbs, with the installation cost being justified due to decreased labor costs alone (the LED bulbs didn't need to be changed as often).

2-5

You need to repair some vertical siding on your house. Each piece of siding is milled to overlap with others, and is 12 inches wide by 8 feet long. You need to replace a section that is 6 feet wide. Each milled board costs $15.50. You also need a gallon of paint ($35.00), sandpaper ($5.50), and a new paint brush ($9.95). Your labor is free. What will this project cost? If you need to replace an additional foot, how much more will it cost?

Solution

Total cost = (6)(15.50) + 35.00 + 5.50 + 9.95 = $143.45

If you need to replace an additional foot, you only need one more board, for an additional $15.50, or $158.95.

2-6

Construction on your custom home has just been completed and the yard must be landscaped. The landscape contractor has estimated that 8 cubic yards of dirt and 3 cubic yards of sand will be required to "level" the yard and provide proper drainage. Dirt is priced at $35 per cubic yard and sand is priced at $27 per cubic yard. She has measured the yard and calculated the area requiring sod to be 3600 ft². Sod is sold in rolls that measure 3' × 30' at a cost of $75 per roll. Shrubbery installation will cost an additional $2500. Estimate the cost of the landscaping.

Solution

Cost of dirt	$8 \times \$35$	=	280
Cost of sand	$3 \times \$27$	=	81
Cost of sod	$3600 \div (3 \times 30) \times 75$	=	3000
Shrubbery installation		=	2500
	Total cost		$5861

2-7

The following data concern one of the most popular products of XYZ Manufacturing. Estimate the selling price per unit.

Labor	= 12.8 hours at $18.75/hour
Factory overhead	= 92% of labor
Material costs	= $65.10
Packing cost	= 10% of materials
Sales commission	= 10% of selling price
Profit	= 22% of selling price

Solution

Labor cost	=	12.8×18.75	=	$240.00
Factory overhead	=	92% of labor	=	220.80
Material cost	=		=	65.10
Packing cost	=	10% of material costs	=	6.51
				$532.41

Let X be the selling price

$$0.10X + 0.22X + 532.41 = X$$
$$0.68X = 532.41$$
$$X = 532.41/0.68 = \$782.96$$

2-8

A tablet computer is being designed with the following estimated costs

Material costs	$16.50
Labor costs	6.00
Overhead	15.00

What wholesale price needs to be charged if they want to make a gross profit of 40%?

Solution

Cost of goods = 16.50 + 6.00 + 15.00 = 37.50
If gross profit = 40%, then cost of goods = 60%
$\quad$ 37.50/price = 0.60
$\quad$ price = 37.50/0.60) = $62.50

2-9

A new 21-kW power substation was built in 2016 in Gibson County for $1.4 million. Weakley County, which is nearby, is planning to build a similar though smaller (18-kW) substation in 2019. The inflation rate has averaged 1.5% per year. If the power sizing exponent is .85 for this type of facility what is the estimated cost of construction of the Weakley County substation?

Solution

Cost of the 21-kW substation in 2019 dollars = 1,400,000(1.015)^3 = $1,463,950

$$C_x = C_k \left(\frac{S_x}{S_k}\right)^n$$

$$C_{18} = C_{21} \left(\frac{18}{21}\right)^{0.85} = 1,463,950(0.8772) = \$1,284,177$$

2-10

American Petroleum (AP) recently completed construction on a large refinery in Texas. The final construction cost was $27,500,000. The refinery covers a total of 340 acres. AP is currently working on plans for a new refinery for Oklahoma. The anticipated size is approximately 260 acres. If the power-sizing exponent for this type of facility is .67, what is the estimated cost of construction?

Solution

$$C_x = C_k \left(\frac{S_x}{S_k}\right)^n$$

$$C_{260} = C_{340} \left(\frac{260}{340}\right)^{0.67} = 27,500,000(0.83549) = \$22,975,975$$

2-11

The time required to produce the first gizmo is 1500 blips. Determine the time required to produce the 450th gizmo if the learning-curve coefficient is .85.

Solution

$$T_i = T_1 \Theta^{(\ln i/\ln 2)}$$
$$T_{450} = 1500(.85)^{(\ln 450/\ln 2)} = 358.1 \text{ blips}$$

2-12

A new training program at Arid Industries is intended to lower the learning-curve coefficient of a certain molding operation that currently costs $95.50/hour. The current coefficient is .87, and the program hopes to lower the coefficient by 10%. Assume that the time to mold the first product is 8 hours. If the program is successful, what cost savings will be realized when the 2000th piece is produced?

Solution

$$T_i = T_1 \Theta^{(\ln i/\ln 2)}$$

Without the training program:
$$T_{2000} = 8(.87)^{(\ln 2000/\ln 2)} = 1.74 \text{ hours}$$
With the training program:
$$T_{2000} = 8(.783)^{(\ln 2000/\ln 2)} = 0.547 \text{ hour}$$
Cost savings $= (1.74 - 0.547)(95.50) = \$113.93$

2-13

Four operations are required to produce a certain product produced by ABC Manufacturing. Use the following information to determine the labor cost of producing the 1000th piece.

	Time Required for 1st Piece	Learning-Curve Coefficient	Labor Cost per Hour
Operation 1	1 hour, 15 minutes	.90	$ 8.50
Operation 2	2 hours	.82	12.00
Operation 3	2 hours, 45 minutes	.98	7.75
Operation 4	4 hours, 10 minutes	.74	10.50

Solution

$T_i = T_1\Theta^{(\ln i/\ln 2)}$

Operation 1:

$T_{1000} = 75(.90)^{(\ln 1000/\ln 2)} = 26.25$ minutes

Cost $= 26.25/60 \times 8.50 = \$3.72$

Operation 2:

$T_{1000} = 120(.82)^{(\ln 1000/\ln 2)} = 16.61$ minutes

Cost $= 16.61/60 \times 12.00 = \$3.32$

Operation 3:

$T_{1000} = 165(.98)^{(\ln 1000/\ln 2)} = 134.91$ minutes

Cost $= 134.91/60 \times 7.75 = \$17.43$

Operation 4:

$T_{1000} = 250(.74)^{(\ln 1000/\ln 2)} = 12.44$ minutes

Cost $= 12.44/60 \times 10.50 = \$2.18$

Total cost $= 3.72 + 3.32 + 17.43 + 2.18 = \$26.65$

2-14

Draw a cash flow diagram for the following end-of-period cash flows.

EOP	Cash Flow
0	−$1000
1	200
2	−100
3	300
4	400
5	−400
6	500

Solution

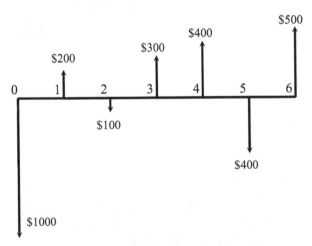

Chapter 3

Interest and Equivalence

3-1

John Buck opens a savings account by depositing $1000. The account pays 6% simple interest. After 3 years John makes another deposit, this time for $2000. Determine the amount in the account when John withdraws the money 8 years after the first deposit.

Solution

F with simple interest $= P + P(i)(n)$

$F = 1000 + 1000(0.06)(8) \leftarrow$ Initial deposit remains in the account for the full 8 years

$\quad + 2000 + 2000(0.06)(5) \leftarrow$ Second deposit remains in the account for 5 years

$F = \$4080$

3-2

If you had $1000 now and invested it at 4%, how much would it be worth 12 years from now?

Solution

$F = 1000(F/P, 4\%, 12) = \$1061.00$

3-3

Mr. Ahmed deposited $200,000 in the Old Third National Bank. If the bank pays 3% interest, how much will he have in the account at the end of 10 years?

Solution

$F = 200,000(F/P, 3\%, 10) = \$268,800$

3-4

If you can earn 2% interest on your money, how much would $1000 paid to you 12 years in the future be worth to you now?

Solution

$P = 1000(P/F, 2\%, 12) = \$788.5$

3-5

Deposited into an account that pays interest monthly, $1000 is allowed to remain in the account for 3 years. Calculate the balance at the end of the 3 years if the annual interest rate is 6%.

Solution

$$i = 6/12 = \frac{1}{2}\% \qquad\qquad n = 12 \times 3 = 36$$

$$F = P(1 + i)^n = 1000(1.005)^{36} = \$1196.68$$

or, using interest tables,

$$F = 1000(F/P, \frac{1}{2}\%, 36) = 1000(1.197) = \$1197$$

3-6

On July 1 and September 1, Abby placed $2000 into an account paying 3% compounded monthly. How much was in the account on October 1?

Solution

$$i = 3/12 = \frac{1}{4}\%$$

$$F = 2000(1 + 0.0025)^3 + 2000(1 + 0.0025)^1 = \$4020.04$$

or

$$F = 2000(F/P, \frac{1}{4}\%, 3) + 2000(F/P, \frac{1}{4}\%, 1) = \$4022.00$$

3-7

Determine the amount of money accumulated in 5 years with an initial deposit of $10,000, if the account earned 12% compounded monthly the first 3 years and 15% compounded semiannually the last 2 years.

Solution

$$F = [10,000(F/P, 1\%, 36)](F/P, 7.5\%, 4)$$

$$= 10,000(1.431)(1.075)^4$$

$$= \$19,110.56$$

3-8

An investment of $10,000 six years ago has now grown to $20,000. Determine the annual interest rate on this investment, assuming annual compounding.

Solution

$$F = P(1 + i)^n$$

$$20,000 = 10,000(1 + i)^6$$

$$(1 + i)^6 = \frac{20,000}{10,000} = 2.0$$

$$1 + i = \sqrt[6]{2.0} = 1.12$$

$$i = 1.12 - 1 = 0.12 = 12\%$$

3-9

Four hundred dollars is deposited into an account that compounds interest quarterly. After 10 quarters the account balance is $441.85. Determine the nominal interest paid on the account.

Solution

$F = P(1 + i)^n$

$41.85 = 400(1 + i)^{10}$

$1.10463 = (1 + i)^{10}$

$1 + i = \sqrt[10]{1.10463} = 1.0100$

$i = 1.0100 - 1 = 0.010 = 1.0\%$

$r = im$

$r = 1.0\%(4) = 4.0\%$

3-10

Margaret M. withdrew $630,315 from an account into which she had invested $350,000. If the account paid interest at 4% per year, she kept the money in the account for how many years?

Solution

$F = P(1 + i)^n$

$630,315 = 350,000(1 + 0.04)^n$

$1.80 = 1.04^n$

$n = ln\dfrac{1.80}{1.04} = 14.99$ years

3-11

Downtown has been experiencing an explosive population growth of 10% per year. At the end of 2017 the population was 16,000. If the growth rate continues unabated, how many years will it take the population to triple?

Solution

Use $i = 10\%$ to represent the growth rate.

$48,000 = 16,000(F/P, 10\%, n)$

$(F/P, 10\%, n) = 48,000/16,000 = 3.000$

From the 10% table, n is 12

Note that population would not have tripled after 11 years.

3-12

If the interest rate is 6% compounded quarterly, how long (number of quarters) will it take to earn $100 interest on an initial deposit of $300?

Solution

$i = 6\%/4 = 1\frac{1}{2}\%$

$400 = 300(F/P, 1\frac{1}{2}\%, n)$

$(F/P, 1\frac{1}{2}\%, n) = 400/300 = 1.333$

From the $1\frac{1}{2}\%$ table, $n = 20$ quarters

3-13

Two years ago, Luckett Land Developers Inc. borrowed $350,000 at a nominal interest rate of 4% compounded quarterly. Due to an economic slowdown, Luckett will be unable to pay off the loan, which is due today. Johnson City Bank has agreed to refinance the loan amount due, plus another $100,000 at a nominal interest rate of 3% compounded monthly. The new loan must be paid off 2 years from now. How much will Luckett owe when the new loan must be paid off?

Solution

Due today $350,000(1 + .01)^8 = \$379,000$

New loan $= 379,000 + 100,000 = \$479,000$

Loan payoff $= 479,000(1 + .0025)^{24} = \$508,582$

3-14

One thousand dollars, deposited into an account that pays interest monthly, is allowed to remain in the account for 3 years. The balance at the end of the 3 years is $1309.00. Calculate the nominal interest paid on this account.

Solution

$F = P(1 + i)^n$

$1309 = 1000(1 + i)^{36}$

$1.309 = (1 + i)^{36}$

$i = 0.75\%$ (per month)

$r = im = 0.75 \times 12 = 9\%$

3-15

Isabella started saving for her retirement 15 years ago. If she invested $30,000 in a stock fund that averaged a 15% rate of return over the 15-year period, and expects to make no more investments and average a 9% return in the future, how long will it be before she has $1,000,000 in her retirement account?

Solution

Value of fund today $F = 30,000(F/P, 15\%, 15) = \$244,110$

$1,000,000 = 244,110(1 + .09)^n$

$4.0965 = (1 + .09)^n$

$n = 16.36 \text{ years} \approx 17 \text{ years}$

3-16

A man decides to put $1000 per month beginning 1 month from today into an account paying 3% compounded monthly. Determine how much (to the nearest penny) will be in the account immediately after the fourth deposit; use only basic concepts.

Solution

Month	Beginning Balance	Interest @ ¼%	Deposit	Ending Balance	
1	$ 0.00	0.00	$1000	$1000.00	
2	1000.00	2.50	1000	2002.50	
3	2002.50	5.01	1000	3007.51	
4	3007.51	7.52	1000	4015.03	←Answer

3-17

Determine the value of P using the appropriate factor.

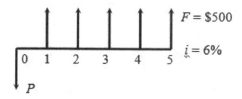

Solution

$P = F(P/F, 6\%, 5) = \$500(0.7473) = \$373.65$

3-18

The Block Concrete Company borrowed $20,000 at 8% interest, compounded semiannually, to be paid off in one payment at the end of 4 years. At the end of the 4 years, Block made a payment of $8000 and refinanced the remaining balance at 6% interest, compounded monthly, to be paid at the end of 2 years. Calculate the amount Block owes at the end of the 2 years.

Solution

$$i_1 = 8/2 = 4\% \qquad n_1 = (4)(2) = 8 \qquad i_2 = 6/12 = \frac{1}{2}\% \quad n_2 = (12)(2)$$
$$F = [20,000(F/P, 4\%, 8) - 8000](F/P, \frac{1}{2}\%, 24) = \$21,841.26$$

3-19

The multistate Powerball Lottery, worth $182 million, was won by a single individual who had purchased five tickets at $1 each. The winner was given two choices: receive 26 payments of $7 million each, with the first payment to be made now and the rest to be made at the end of each of the next 25 years, or receive a single lump-sum payment now that would be equivalent to the 26 payments of $7 million each. If the state uses an interest rate of 4% per year, find the amount of the lump sum payment.

Solution

$$P = 7,000,000 + 7,000,000(P/F, 4\%, 25) = \$116,354,000$$

3-20

A woman deposited $10,000 into an account at her credit union. The money was left on deposit for 10 years. During the first 5 years, the woman earned 9% interest, compounded monthly. The credit union then changed its interest policy; as a result, in the second 5 years the woman earned 6% interest, compounded quarterly.

a.   How much money was in the account at the end of the 10 years?
b.   Calculate the rate of return that the woman received.

Solution

a.   At the end of 5 years:
$$F = 10,000 \ (F/P, \frac{3}{4}\%, 60)* = \$15,660.00 \qquad * \ i = 9/12 = \frac{3}{4}\% \qquad n = (12)(5) = 60$$
At the end of 10 years:
$$F = 15,660(F/P, 1\frac{1}{2}\%, 20)** = \$21,094.02 \quad ** \ i = 6/4 = 1\frac{1}{2}\% \qquad n = (4)(5) = 20$$
b. $10,000(F/P, i, 10) = 21,094.02$
$(F/P, i, 10) = 2.1094$
Try $i = 7\%$ $\qquad (F/P, 7\%, 10) = 1.967$
Try $i = 8\%$ $\qquad (F/P, 8\%, 10) = 2.159$
Interpolating, $i = 7.75\%$

3-21

Lexie C. deposits $4000 into an account paying 3% simple interest. After 3.5 years, she moves the money into an account that will earn 4% compounded quarterly. After another 4 years, she deposits another $2000 into the account. How much will be in the account 1 year after the final deposit?

Solution

$F = P + P(i)(n)$ (with simple interest)

$F = 4000 + 4000(0.03)(3.5) = \$4420$

$F = P(F/P\ i,\ n)$ (with compound interest)

$F = 4420(F/P\ 1\%,\ 16) = \$5184.66$

After last deposit money remains in account for one more year

$F = P(F/P\ i,\ n)$

$F = (2000 + 5184.66)(F/P\ 1\%,\ 4) = \$7479$

3-22

Justin P. deposited \$2000 into an account 5 years ago. Simple interest was paid on the account. He has just withdrawn \$2876. What interest rate did he earn on the account?

Solution

$F = P + P(i)(n)$ (with simple interest)

$2876 = 2000 + 2000(i)(5)$

$876 = 10,000i$

$i = 0.0876 = 8.76\%$

3-23

A sum of \$5000 is invested for 5 years with annual interest rates of 9% the first, 8% the second, 12% the third, 6% the fourth, and 15% the fifth years, respectively. Determine the future amount after 5 years.

Solution

$F = P(1 + i)^n$

$F = (((((5000(1 + 0.09)^1)(1 + 0.08)^1)\ (1 + 0.12)^1)(1 + 0.06)^1)(1 + 0.15)^1)$

$\quad = 5000(1.09)(1.08)(1.12)(1.06)(1.15)$

$\quad = \$8036.04$

3-24

Callis Construction LLC is planning to expand its site planning division 4 years from now. In order to expand, the CFO of the company has determined that \$400,000 will be required. Eighth National Bank has agreed to pay 8% simple interest for the first 2 years the money is on deposit with the terms changing to a 6% nominal rate compounded monthly for the last 2 years. How much should Callis deposit today in order to have the needed \$400,000?

Solution

$F_4 = \$400,000$

Two years from now the amount required to yield $400,000 at end of Year 4

$P_2 = 400,000(1 + .005)^{-24} = \$354,874.27$

Today $P = 354,874.27(1 + .08(2)) = \$305,926$

3-25

Five years ago Mary Skinner inherited a sum of $50,000 and bought 1000 shares of IBEM at $50 per share. She sold 400 shares of the stock 3 years ago for $57 per share and invested the money in a CD that paid 2% interest per year compounded quarterly. Mary cashed the CD today and sold the remaining 600 shares of stock for $65 per share. What nominal interest rate did she earn on her inheritance?

Solution

Proceeds from past stock sale $= 400 \times 57 = \$22,800$

CD value today $= 22,800(F/P, \frac{1}{2}\%, 12) = \$24,213.60$

Proceeds from current stock sale $= 600 \times 65 = \$39,000$

Total amount of money $= 24,213.60 + 39,000 = \$63,213.60$

Calculating nominal interest rate:

$F = P(1+ i)^n$

$63,213.60 = 50,000(1+ i)^{20}$

$1.2643 = (1+ i)^{20}$

$i = 0.0118 = 1.18\%$

$r = im = 1.18 \times 4 = 4.72\%$

3-26

A local bank is advertising to savers a rate of 6% compounded monthly, yielding an effective annual rate of 6.168%. If $2000 is placed in savings now and no withdrawals are made, how much interest (to the penny) will be earned in 1 year?

Solution

Interest = Effective annual rate × Principal $= 0.06168 \times 2000 = \$123.36$

Monthly compounding is irrelevant when the effective rate is known.

3-27

A small company borrowed $10,000 to expand the business. The entire principal of $10,000 will be repaid in 2 years, but quarterly interest of $330 must be paid every 3 months. What nominal annual interest rate is the company paying?

Solution

The $330 is interest for one period; therefore $i = 330/10{,}000 = 3.3\%$ per quarter:

$r = 3.3 \times 4 = 13.2\%$ nominal annual rate

3-28

Cole's Home Solutions policy is to charge 2¼% interest each month on unpaid credit balances. What nominal interest is Cole's charging? What is the effective interest?

Solution

(a) $r = im = 12 \times 2.25 = 27\%$

(b) $i_{\text{eff}} = (1 + i)^n - 1 = (1.0225)^{12} - 1 = 0.3060 = 30.60\%$

3-29

EZ Pay Loans will lend you $100 today with repayment of $117.50 due in 1 month. Determine the nominal and effective rate of this loan.

Solution

Interest $= \$117.50 - 100 = \$17.50$ for 1 month

$i = 17.50/100 = 17.5\%$

$r = im = 17.5 \times 12 = 210\%$

$i_{\text{eff}} = (1 + i)^n - 1 = (1.175)^{12} - 1 = 5.9256 = 592.56\%$

3-30

For a nominal interest rate of 6%, what is the effective interest rate if interest is

a.　　　compounded quarterly?

b.　　　compounded monthly?

c.　　　compounded continuously?

Solution

a.　$i_{\text{eff}} = [(1 + 0.015)^4 - 1] = 0.0614 = 6.14\%$

b.　$i_{\text{eff}} = [(1 + 0.005)^{12} - 1] = 0.0617 = 6.17\%$

c.　$i_{\text{eff}} = [e^{.06} - 1] = 0.0618 = 6.18\%$

3-31

Which is the better investment, a fund that pays 5% compounded annually or one that pays 4.8% compounded continuously?

Solution

i_{eff} = 5% compounded annually = $[(1 + 0.5)^1 - 1] = 0.05 = 5\%$

i_{eff} = 4.8% compounded continuously = $[e^{.048} - 1] = 0.0492 = 4.92\%$

Therefore, 5% compounded annually is slightly better.

3-32

The effective interest rate is 9.38%. If interest is compounded monthly, what is the nominal interest rate?

Solution

$i_{\text{eff}} = (1 + (r/m))^m - 1 \Rightarrow r/m = (1 + i_{\text{eff}})^{1/m} - 1 = (1.0938)^{1/12} - 1 = 0.0075 = 0.75\%$

$r = 12 \times 0.75 = 9\%$

3-33

The effective interest rate on a mortgage with monthly payments is 9.38%. What is the monthly interest rate on the mortgage? What is the nominal interest rate?

Solution

$i_{\text{eff}} = [(1 + i)^m - 1] = 9.38\%$

$0.0938 = [(1 + i)^{12} - 1]$

$1.0938 = (1 + i)^{12}$

$i = 0.75\%$ (per month)

$r = im = 0.75 \times 12 = 9\%$

Chapter 4

Equivalence for Repeated Cash Flows

4-1

Dylan deposits $10,000 now and makes an additional deposit of $5,000 at the end of Year 3 in the same account. Determine the balance in the account at the end of 10 years if the interest rate is a nominal 8%, compounded quarterly.

Solution

$F = 10,000(F/P, 2\%, 40) + 5000(F/P, 2\%, 28) = \$30,785$

4-2

The Smiths are planning ahead for their daughter's education. She's 8 now and will start college in 10 years. How much will they need to set aside each year to have $65,000 in 10 years if the annual interest rate is 7%?

Solution

$A = 65,000(A/F, 7\%, 10) = 65,000(.0724) = \$4706$

4-3

A young engineer wishes to buy a house but can afford monthly payments of only $500. Thirty-year loans are available at 6% interest compounded monthly. If she can make a $5000 down payment, what is the price of the most expensive house that she can afford to purchase?

Solution

$i = 6\%/12 = \frac{1}{2}\%$ per month $\qquad n = 30 \times 12 = 360$

$P* = 500(P/A, \frac{1}{2}\%, 360) = 83,396.00$

$P = 83,396.00 + 5000$

$P = \$88,396$

4-4

A person borrows $15,000 at an interest rate of 6%, compounded monthly, to be paid with payments of $456.33.

a.   What is the length of the loan in years?

b.   What is the <u>total</u> amount that would be required at the end of the twelfth month to pay off the entire loan balance?

Solution

   a. $P = A(P/A, i\%, n)$

        $15{,}000 = 456.33(P/A, \frac{1}{2}\%, n)$

        $(P/A, \frac{1}{2}\%, n) = 15{,}000/456.33 = 32.871$

        From the $\frac{1}{2}\%$ interest table, $n = 36$ months $= 3$ years.

   b. $P = 120^{th}$ payment $+$ PW of remaining 24 payments

        $= 456.33 + 456.33(P/A, \frac{1}{2}\%, 24)$

        $= \$10{,}752.50$

4-5

A $50,000 loan with a nominal interest rate of 6% is to be repaid over 30 years with payments of $299.77. The borrower wants to know how many payments, N^*, he will have to make until he owes only half of the amount he borrowed initially.

Solution

    The outstanding principal is equal to the present worth of the remaining payments when the payments are discounted at the loan's effective interest rate.

    Therefore, let N' be the remaining payments.

    $\frac{1}{2}(50{,}000) = 299.77(P/A, \frac{1}{2}\%, N')$

    $(P/A, \frac{1}{2}\%, N') = 83.397$

    $N' = 108.30 \approx 108$ From $i = \frac{1}{2}\%$ table

    So, $N^* = 360 - N' = 252$ payments

4-6

While in college, Ellen received $40,000 in student loans at 8% interest. She will graduate in June and will repay the loans in either 5 or 10 equal annual payments. Compute her yearly payments for both repayment plans.

Solution

<u>5 Years</u>	<u>10 Years</u>
$A = P(A/P, i, n)$	$A = P(A/P, i, n)$
$= 40{,}000(A/P, 8\%, 5)$	$= 40{,}000(A/P, 8\%, 10)$
$= \$10{,}020.00$	$= \$5{,}960.00$

4-7

Given:

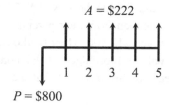

$A = \$222$

1 2 3 4 5

$P = \$800$

Find: i

Solution

$P = A(P/A, i\%, 5)$

$800 = 222(P/A, i\%, 5)$

$(P/A, i\%, 5) = 800/222 = 3.6$

From the interest tables, $i = 12\%$.

4-8

J. D. Homeowner has just bought a house with a 20-year, 9%, $70,000 mortgage on which he is paying $629.81 per month.

a.   If J. D. sells the house after 10 years, how much must he pay the bank to completely pay off the mortgage at the time of the 120th payment?

b.   How much of the first $629.81 payment on the loan is interest?

Solution

a. $P = 120^{th}$ payment + PW of remaining 120 payments

$= 629.81 + 629.81(P/A, \frac{3}{4}\%, 120)$

$= \$50,348.27$

b. $\$70,000 \times 0.0075 = \$525$

4-9

Suppose you wanted to buy a $180,000 house. You have $20,000 cash to use as the down payment. The bank offers to lend you the remainder at 6% nominal interest. The term of the loan is 20 years. Compute your monthly loan payment.

Solution

Amount of loan: $\$180,000 - \$20,000 = \$160,000$

$i = 6\%/12 = \frac{1}{2}\%$ per month $n = 12 \times 20 = 240$

$A = 160,000(A/P, \frac{1}{2}\%, 240) = \$1145.60$ per month

4-10

To offset the cost of buying a $120,000 house, Jose and Sophia borrowed $25,000 from their parents at 6% nominal interest, compounded monthly. The loan from their parents is to be paid off in 5 years in equal monthly payments. The couple has saved $12,500. Their total down payment is therefore $25,000 + 12,500 = $37,500. The balance will be mortgaged at 9% nominal interest, compounded monthly for 30 years. Find the combined monthly payment that the couple will be making for the first 5 years.

Solution

> Payment to parents:
> > $25,000(A/P, \frac{1}{2}\%, 60) = \$482.50$
>
> Borrowed from bank: $120,000 - 37,500 = \$82,500$
> Payment to bank:
> > $82,500(A/P, \frac{3}{4}\%, 360) = \$664.13$
>
> Therefore, monthly payments are $482.50 + 664.13 = \$1,146.63$.

4-11

Martinez and Martinez Inc. makes monthly payments of $152.11 and pays 6% interest on a current loan. The initial loan was scheduled to be paid off in 3 years. Determine the loan amount.

Solution

> $P = 152.11(P/A, \frac{1}{2}\%, 36) = \$5000$

4-12

Abby's cat, Si, has convinced her to set up an account that will assure him of his Meow Mix for the next four years. Abby will deposit an amount today that will allow Si to make end-of-month withdrawals of $10 for the next 48 months. Assume an interest rate of 6% and that the account will have a zero balance when the last withdrawal is made. What is the amount that Abby should deposit? What is the account balance immediately after the 24th withdrawal is made?

Solution

> $P = 10(P/A, \frac{1}{2}\%, 48)$
> > $= \$425.80$
>
> To determine the account balance, calculate the P of the remaining withdrawals.
> $P = 10(P/A, \frac{1}{2}\%, 24) = \$225.63$

Or, determine the future value, given the initial payment and the withdrawals.

	A	B	C	D	E	F	G	H	I
1	Problem	i	N	PMT	PV	FV	Solve for	Answer	Formula
2	4-12	0.50%	24	-10.00	$426		FV	$225.63	=-FV(B2,C2,D2,E2)

The future value is $225.63.

4-13

Ben Spendlove just purchased a new stereo system for $975 and will be making payments of $45 per month. How long will it take to completely pay off the stereo at 18% nominal interest?

Solution

$i = 18\%/12 = 1\frac{1}{2}\%$ per month

$975 = 45(P/A, 1\frac{1}{2}\%, n)$

$(P/A, 1\frac{1}{2}\%, n) = 975/45 = 21.667$

From the $1\frac{1}{2}\%$ table, n is between 26 and 27 months. The loan will not be completely paid off after 26 months. Therefore, the final payment will be less than $45.

4-14

Henry Fuller has agreed to purchase a used automobile for $13,500. He wishes to limit his monthly payment to $350 for a period of 2 years. What down payment must he make to complete the purchase if the interest rate on the loan will be 6%?

Solution

$P = P' + A(P/A, \frac{1}{2}\%, 24)$ where P' is the down payment

$13,500 = P' + 350(22.563)$

$P' = 13,500 - 7897.05 = \$5602.95$

4-15

Beginning 1 month from today David B. will deposit each month $200 into an account paying 6% nominal interest. He will make a total of 240 deposits (20 years). After the last deposit the money in the account will begin to earn 4% interest compounded annually. After another 10 years David will begin to withdraw annual amounts for a total of 10 years. How much can be withdrawn each year if the account is to be depleted (zero balance) after another 10 years?

Solution

$F_{20} = 200(F/A, \frac{1}{2}\%, 240) = 200(462.041) = \$92,408.20$

$F_{30} = 92,408.20(F/A, 4\%, 10) = 92,408.20(1.480) = \$136,764.14$

$A = 136,764.14(A/P, 4\%, 10) = 136,764.14(.1233) = \$16,863.02$

4-16

Ray Witmer, an engineering professor at UTM, is preparing to retire to his farm and care for his cats and dogs. During his many years at UTM he invested well and has a balance of \$1,098,000 in his retirement fund. How long will he be able to withdraw \$100,000 per year, beginning today, if his account earns interest at a rate of 4% per year?

Solution

$A = \$100,000$ \qquad $P = 1,098,000 - 100,000* = \$998,000$ \qquad *First withdrawal is today.

$100,000 = 998,000(A/P, 4\%, n)$

$(A/P, 4\%, n) = 100,000/998,000$

$(A/P, 4\%, n) = 0.1002$

From the $i = 4\%$ table, $n = 13$ additional years of withdrawals, 14 total years of withdrawals.

4-17

Benita deposits \$125 per month into an account paying 6% interest for 2 years, to be used to purchase a car. The car she selects costs more than the amount in the account. She agrees to pay \$175 per month for 2 more years at 9% interest, and also uses a gift from her uncle of \$500 as part of the down payment. What is the cost of the car to the nearest dollar?

Solution

$i = 6\%/12 = \frac{1}{2}\%$ \qquad\qquad $n = 12 \times 2 = 24$

$F = 125(F/A, \frac{1}{2}\%, 24) = 125(24.310) = \$3098.75$ ← Amount in account

$i = 9\%/12 = \frac{3}{4}\%$ \qquad\qquad $n = 12 \times 2 = 24$

$P = 175(P/A, \frac{3}{4}\%, 24) = 175(21.889) = \$3830.58$ ← Amount repaid by loan

Total $= 3098.75 + 3830.58 + 500 = \$7429.33 = \$7429$ ← Cost of automobile

4-18

Lenagar Lumber Inc. is making monthly payments of \$572.39 and paying nominal 9% interest on a current loan. The initial loan was scheduled to be paid off in 3 years. Immediately after the 15th payment business conditions allow Lenagar to pay off the loan. Determine the amount of the loan balance due after the 15th payment.

Solution

36 months − 15 = 21 remaining monthly payments

$P = 572.39(P/A, \frac{3}{4}\%, 21) = 572.39(19.363) = \$11,083$

4-19

Jason W. bought a Mercedes when he came to UTM as an engineering student. The Mercedes was purchased by taking a loan that was to be paid off in 20 equal, quarterly payments. The interest rate on the loan was 12%. Four years later, after Jason made his 16th payment, he got married (no more dating!) and sold the Mercedes to his buddy Houston S. Houston made arrangements with Jason's bank to refinance the loan and to pay Jason's unpaid balance by making 16 equal, quarterly payments at the same interest rate that Jason was paying. Houston flunked out of UTM (too many dates!) 3¼ years later, after having made his 13th payment; he then sold the car to Jeff M. Jeff paid the bank $2000 cash (he had a good summer job!) to pay the loan balance. How much had Jason borrowed to buy the new Mercedes?

Solution

$i = 12\%/4 = 3\%$ per quarter

Jason W.

$A = P(A/P, 3\%, 20)$

$A = P(0.0672)$ Quarterly payment for Jason

Jason owes

$P = 0.0672P(P/A, 3\%, 4)$ Present worth of four remaining payments

$= 0.0672P(3.717)$

$= 0.2498P$

Houston S.

$A = 0.2498P(A/P, 3\%, 16)$ Quarterly payment for Houston

$= 0.2498P(0.0796)$

$= 0.0199P$

Jeff M.

$P = 0.0199P(P/A, 3\%, 3)$ Present worth of three remaining payments

$= 0.0199P(2.829)$

$= 0.0563P$

Set final payment equal to present worth of remaining payments:

$2000 = 0.0562P$

$P = \$35,556.75$

4-20

You have just taken out a mortgage of $50,000 for 30 years, with monthly payments at 6% interest. The same day you close on the mortgage you receive a $25,000 gift from your parents to be applied to the mortgage principal. What amount of time will now be required to pay off the mortgage if you continue to make the original monthly payments? What is the amount of the last payment? (Assume any residual partial payment amount is added to the last payment.)

Solution

$i = 6\%/12 = \frac{1}{2}\%$ per month $\qquad n = 12 \times 30 = 360$ periods (months)

$A = 50,000(A/P, \frac{1}{2}\%, 360) = 50,000(0.00600) =$

$\quad = \$300.00$ monthly payment

After reduction of P to 25,000,

$25,000 = 300.00(P/A, \frac{1}{2}\%, n)$

$(P/A, \frac{1}{2}\%, n) = 83.33$

At $n = 104$ periods: $P/A = 80.942$

At $n = 120$ periods: $P/A = 90.074$

By interpolation, $n = 108.18$ periods $= 9.02$ years.

At 9 years (108 periods): $P = 300.00(P/A, \frac{1}{2}\%, 108)$

$\qquad\qquad\qquad\qquad = 300.00(83.33)$

$\qquad\qquad\qquad\qquad = \$24,999.00$

Residual $= 25,000 - 24,999.00 = \$1.00$

Last payment $=$ Value of residual at time of last payment $+$ Last payment

$\qquad\qquad\qquad = 1.00 + 300.00 = \$301.00$

4-21

Mr. Deere just purchased a new riding lawn mower. The first year maintenance is free. His maintenance costs are estimated to be $15 the second year and increase by $15 each year thereafter. At 8% interest how much money should Mr. Deere set aside when he buys the mower to pay for maintenance for the next 10 years?

Solution

$P = 0(P/A, 8\%, 10) + 15(P/G, 8\%, 10) = \$389.66$

4-22

Today Sam Keel deposits $5,000 in an account that earns 4% compounded quarterly. Additional deposits are made at the end of each quarter for the next 10 years. The deposits start at $100 and increase by $50 each quarter thereafter. Determine the amount that has accumulated in the account at the end of 10 years.

Solution

$$F = 5000(F/P, 1\%, 40) + [100 + 50(A/G, 1\%, 40)] (F/A, 1\%, 40) = \$56,766$$

4-23

Find the uniform annual equivalent for the following cash flow diagram if $i = 10\%$. Use the appropriate gradient and uniform series factors.

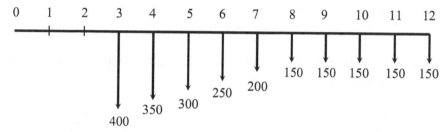

Solution

$$P_1 = [400(P/A, 10\%, 6) - 50(P/G, 10\%, 6)](P/F, 10\%, 2) = \$1039.45$$
$$P_2 = [150(P/A, 10\%, 4)](P/F, 10\%, 8) = \$221.82$$
$$P = 1039.45 + 221.82 = \$1261.27$$
$$A = 1261.27(A/P, 10\%, 12) = \$185.15$$

4-24

Find the present equivalent of the following cash flow diagram if $i = 8\%$.

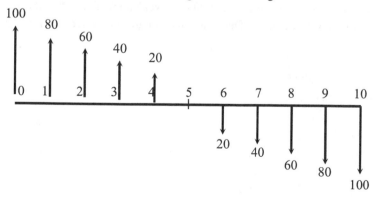

Solution

$$P = 100 + 80(P/A, 8\%, 10) - 20(P/G, 8\%, 10) = \$117.26$$

4-25

Holloman Hops has budgeted $300,000 per year to pay for labor over the next 5 years. If the company expects the cost of labor to increase by $10,000 each year, and the interest rate is 10%, what is the expected cost of the labor in the first year?

Solution

$A' = \$300,000$ the budgeted amount

$A' = A + 10,000(A/G, 10\%, 5)$

$300,000 = A + 10,000(1.81)$

$A = \$281,900$ first-year labor cost

4-26

For the cash flow shown, determine the value of G that will make the future worth at the end of Year 6 equal to $8000 at an interest rate of 6% per year.

Year	0	1	2	3	4	5	6
Cash Flow	0	600	$600 + G$	$600 + 2G$	$600 + 3G$	$600 + 4G$	$600 + 5G$

Solution

$$P = 8000(P/F, 6\%, 6) = 8000(0.7050) = \$5640$$

$$5640 = 600(P/A, 6\%, 6) + G(P/G, 6\%, 6)$$

$$5640 = 600(4.917) + G(11.459)$$

$$G = \$234.73$$

4-27

Deposits are made at the end of Years 1 through 7 into an account paying 3%. The first deposit equals $5000 and each deposit will increase by $1000 each year thereafter. After the last deposit assume no deposits or withdrawals are made. Determine the amount in the account after 10 years.

Solution

$$PV = 5000(P/A, 3\%, 7) + 1000(P/G, 3\%, 7)$$

$$= 5000(6.230) + 1000(17.955)$$

$$= \$49,105$$

$$FV = 49,105(F/P, 3\%, 10)$$

$$= 43,360(1.344)$$

$$= \$65,997.12$$

4-28

Determine the uniform annual equivalent of the following cash flows at an interest rate of 6%.

t	0	1	2	3	4	5	6	7	8	9	10
CF_t	$100	$90	$80	$70	$60	$50	$50	$60	$70	$80	$90

Solution

$P = 100 + 90(P/A, 6\%, 5) - 10(P/G, 6\%, 5) + [50(P/A, 6\%, 5) + 10(P/G, 6\%, 5)]$

$\quad = \$616.33$

$A = 616.33(A/P, 6\%, 10)$

$\quad = \$83.76$

4-29

A set of cash flows begins at \$25,000 the first year, increasing 10% each year until n = 10 years. If the interest rate is 5%, what is the present value?

Solution

▲	A	B
1	5%	interest rate
2	Year	Cash Flow
3	1	25,000
4	2	27,500
5	3	30,250
6	4	33,275
7	5	36,603
8	6	40,263
9	7	44,289
10	8	48,718
11	9	53,590
12	10	58,949
13	NPV	$296,166
14		=NPV(A1,B3:B12)

The present value is \$296,166.

4-30

A set of cash flows begins at \$25,000 the first year, decreasing 10% each year until n = 10 years. If the interest rate is 4%, what is the present value?

Solution

▲	A	B
1	4%	interest rate
2	Year	Cash Flow
3	1	25,000
4	2	22,500
5	3	20,250
6	4	18,225
7	5	16,403
8	6	14,762
9	7	13,286
10	8	11,957
11	9	10,762
12	10	9,686
13	NPV	$136,508
14		=NPV(A1,B3:B12)

The present value is $136,508.

4-31

Maintenance costs on some mining equipment are expected to be $20,000 the first year, increasing 10% each year until the equipment is retired after 4 years. If the interest rate is 12%, what is the net present value of the maintenance costs.

Solution

$$PW = \$20,000 \, [\, 1 - (1 + 0.10)^4 \, (1 + 0.12)^{-4} \,] / (0.12 - 0.10)$$
$$PW = \$20,000 \, (3.48) = \$69,600$$

4-32

How much will accumulate in an Individual Retirement Account (IRA) in 15 years if $5000 is deposited in the account at the end of each quarter during that time? The account earns 4% interest, compounded quarterly. What is the effective interest rate?

Solution

$i = 4\%/4 = 1\%$ per quarter $n = 4 \times 15 = 60$

$F = 5000 \, (F/A, 1\%, 60) = \$408,350$

Effective interest rate $= (1 + 0.01)^4 - 1 = 4.06\%$

4-33

Patrick J. just purchased a Zündapp Janus (a model and brand of car) for $4000. He agreed to pay for the car with monthly payments of $138.80 over a 36-month period. What nominal interest rate is he paying on the loan? What is the effective rate?

Solution

$P = \$4000 \quad A = \$138.80$

$138.80 = 4000(A/P, i\%, 36)$

$0.0347 = (A/P, i\%, 36)$

Searching the interest tables where $n = 36$ for $A/P = 0.0347$ determines $i = 1\frac{1}{4}\%$

$r = im$

$r = 1\frac{1}{4}\% \times 12 = 15\%$ nominal annual

$i_{eff} = (1 + (r/m))^m - 1 \qquad\qquad r/m = i$

$i_{eff} = (1 + 0.0125)^{12} - 1 = 0.1608 = 16.08\%$

4-34

A bank is offering a loan of $20,000 with an interest rate of 9%, payable with monthly payments over a 4-year period.
a. Calculate the monthly payment required to repay the loan.
b. This bank also charges a loan fee of 4% of the amount of the loan, payable at the time of the closing of the loan (that is, at the time the borrower receives the money). What effective interest rate is the bank charging?

Solution

a. The monthly payments:

$i = 9\% /12 = \frac{3}{4}\%$ monthly, $\qquad n = 12 \times 4 = 48$ months

$20,000(A/P, \frac{3}{4}\%, 48) = \$498$

b. Actual money received $= P = 20,000 - 0.04(20,000) = \$19,200$

$A = \$498$ based on $20,000

Recalling that $A = P(A/P, i, n)$

$498 = 19,200(A/P, i, 48)$

$(A/P, i, 48) = 498/19,200 = 0.0259$

For $i = \frac{3}{4}\%$, the A/P factor at $n = 48 = 0.0249$

For $i = 1\%$, the A/P factor at $n = 48 = 0.0263$

By interpolation, $\qquad i \approx \frac{3}{4} + \frac{1}{4}[(0.0259 - 0.0249)/(0.0263 - 0.0249)] \qquad i \approx 0.929\%$

Therefore, $i_{eff} = (1 + 0.00929)^{12} - 1 = 0.117 = 11.7\%$.

4-35

The effective interest rate on a mortgage with monthly payments is 9.38%. What is the monthly payment if the original mortgage amount is $200,000 and the mortgage is to be paid over 30 years?

Solution

$$i_{eff} = (1 + (r/m))^m - 1 \qquad r/m = i$$
$$i_{eff} = (1 + i)^m - 1$$
$$0.0938 = (1 + i)^{12} - 1$$
$$1.0938 = (1 + i)^{12}$$
$$i = .0075 = .75\%$$

$$A = 200,000(A/P, \tfrac{3}{4}\%, 360) = \$1610$$

Using a spreadsheet,

◢	A	B	C	D
1	9.38%	effective rate		
2	$200,000	Principal		
3	30	years		
4	360	months		
5	Nominal:	9.00%	=NOMINAL(A1,12)	
6	Monthly:	0.75%	=B5/12	
7				
8	Payment:	$1,609.15	=PMT(B6,A4,-A2)	

4-36

Big John Sipes, owner of Sipes's Sipping Shine, has decided to replace the distillation system his company now uses. After some research, he finds an acceptable distiller that costs $62,500. The current machine has approximately 1200 lbs. of copper tubing that can be salvaged and sold for $4.75/lb. to use as a down payment on the new machine. The remaining components of the distillation machine can be sold as scrap for $3000. This amount will also be used to pay for the replacement equipment. The remaining money will be obtained through a 10-year mortgage with quarterly payments at an interest rate of 8%. Determine the quarterly payment required to pay off the mortgage. Also determine the effective interest rate on the loan.

Solution

$$i = 8/4 = 2\% \quad n = 4 \times 10 = 40$$
$$P = 62,500 - (1,200 \times 4.75) - 3000 = 53,800$$
$$A = 53,800(A/P, 2\%, 40)$$
$$= 53,800(0.0366)$$
$$= \$1969$$
$$i_{eff} = (1 + 0.02)^4 - 1 = 8.24\%$$

4-37

Twelve monthly payments are needed to pay off a $2000 loan. The annual interest rate is 12.5%. Build a table that shows the balance due, principal payment, and the interest payment for each month. How much interest was paid?

Solution

	A	B	C	D	E
1	2,000	Loan amount			
2	13%	Annual interest rate			
3	1.042%	Period interest rate			
4	12	# payments			
5					
6	$178.17	$ payment			
7					
8	Month	Principal Owed	Interest Owed	Monthly Payment	Principal Paid
9	1	2000.00	20.83	$178.17	157.33
10	2	1842.67	19.19	$178.17	158.97
11	3	1683.70	17.54	$178.17	160.63
12	4	1523.07	15.87	$178.17	162.30
13	5	1360.77	14.17	$178.17	163.99
14	6	1196.78	12.47	$178.17	165.70
15	7	1031.08	10.74	$178.17	167.43
16	8	863.65	9.00	$178.17	169.17
17	9	694.48	7.23	$178.17	170.93
18	10	523.55	5.45	$178.17	172.71
19	11	350.84	3.65	$178.17	174.51
20	12	176.33	1.84	$178.17	176.33
21					
22	Total		$137.99	$2,137.99	$2,000.00

$137.99 was paid in interest.

4-38

Determine the present value of cash flows that start at $25,000 and increase 4% per year, ending in year 10. The interest rate is 5%.

Solution

◢	A	B	C
1	$25,000	Initial	
2	4%	g	
3	5%	i	
4			
5	Year	Cash Flows	
6	1	25000	
7	2	26000	=+B6*(1+A2)
8	3	27040	
9	4	28122	
10	5	29246	
11	6	30416	
12	7	31633	
13	8	32898	
14	9	34214	
15	10	35583	
16	NPV	$228,146	=NPV(A3,B6:B15)

The present value is $228,146.

4-39

Find the *annual* worth of a quarterly lease payment of $500 at 8% interest.

Solution

Lease payments are beginning-of-period cash flows.
First find the present worth of the quarterly payments at 8/4 = 2%.
$P = 500 + 500(P/A, 2\%, 3) = \$1941.95$
$A = 1941.95(1 + 0.02)^4 = \$2102$

4-40

A person would like to retire 15 years from now. He currently has $132,000 in savings, and he plans to deposit $800 per month, starting next month, in a special retirement plan. The $132,000 is earning 8% interest, while the monthly deposits will pay him 6% annual interest. Once he retires, he will deposit the total of the two sums of money into an account that he expects will earn a 4% annual interest rate. Assuming that he will spend only the interest he earns, how much will he collect in annual interest, starting in Year 16?

Solution

Savings: $F = 132,000(F/P, 8\%, 15) = \$551,364$
Monthly deposits: $F = 800(F/A, \frac{1}{2}\%, 180) = \$232,655$
The total amount on deposit at the end of Year 10 is
$F_{Total} = 551,364 + 232,655 = \$784,019$
Interest per year = $784,019 \times 0.04 = \$31,361$

Chapter 5

Present Worth Analysis

5-1

A tax refund expected 1 year from now has a present worth of $3000 if $i = 6\%$. What is its present worth if $i = 4\%$?

Solution

The actual value of the refund will be $3000(F/P, 6\%, 1) = 3000(1.06) = \$3180$

The PW if i = 4% is $3180(P/F, 4\%, 1) = 3180(0.9615) = \$3057.57$

5-2

Investment in a crane is expected to produce profit from its rental of $15,000 during the first year of service. The profit is expected to decrease by $2500 each year thereafter. At 12% interest, find the present worth of the profits.

Solution

$$P = 15,000(P/A, 12\%, 6) - 2500(P/G, 12\%, 6) = \$39,340$$

5-3

Strickland Storage Inc. leases storage units for $200/month. Calculate the present worth of 12 lease payments at 6%.

Solution

Leases are beginning-of-period payments.

PW = $200 + 200(P/A, \frac{1}{2}\%, 11)$

= $2335.40

Using a spreadsheet,

	A	B	C	D	E	F	G	H	I
1	Problem	i	N	PMT	PV	FV	Solve for	Answer	Formula
2	5-3	0.5%	12	-200		0	PV	$2,335.41	=PV(B2,C2,D2,F2,1)

PW = $2,335.41. Note the use of a '1' for 'type' in the formula.

5-4

The winner of a sweepstakes prize is given the choice of a one-time payment of $1,000,000 or a guaranteed $80,000 per year for 20 years. If the value of money is 5%, which option should the winner choose?

Solution

Option 1: $P = \$1,000,000$
Option 2: $P = 80,000(P/A, 5\%, 20) = \$996,960$
Choose Option 1: take the $1,000,000 now.

5-5

A local car wash charges $3 per wash, or one can pay $12.98 for 5 washes, payable in advance with the first wash. If you normally washed your car once a month, and your cost of money is 1% compounded monthly, would the option be worthwhile?

Solution

$NPV_{Pay\ for\ 5} = -\$12.98$
$NPV_{Pay/Wash} = -3.00 - 3.00(P/A, 1\%, 4) = -\$14.71$
The "pay for 5" option is the more economical.

5-6

A project being considered by the XYZ Company will have $100,000 in construction costs in each of the first 3 years of the project. Income of $100,000 will begin flowing in Year 4 and will continue through Year 10. Find the net present worth at 4% of the project

Solution

$P = -100,000(P/A, 4\%, 3) + 100,000(P/A, 4\%, 7)(P/F, 4\%, 3)$
$\quad = \$256,077.80$

5-7

The community theater spends $10,000 annually to produce a musical extravaganza. Immediately *before* this year's extravaganza, the theater had a balance of $60,000 in an account paying 4% interest. *After* this year, how many extravaganzas can be sponsored without raising more money?

Solution

$60,000 - 10,000 = 10,000(P/A, 4\%, n)$
$(P/A, 8\%, n) = 50,000/10,000 = 5$
From the $i = 4\%$ table, $n = 5$.
So 5 is the number of extravaganza after this year's.
There will be some money left over but not enough to pay for a 6th year.

5-8

The annual income from an apartment house is $20,000. The annual expense is estimated to be $2000. If the apartment house can be sold for $100,000 at the end of 10 years, how much should you be willing to pay for it now, with a required return of 10%?

Solution

$$P = (A_{INCOME} - A_{EXPENSE})(P/A, 10\%, 10) + F_{SALE}(P/F, 10\%, 10)$$
$$= (20,000 - 2,000)(P/A, 10\%, 10) + 100,000(P/F, 10\%, 10)$$
$$= \$149,160$$

Using a spreadsheet,

◢	A	B	C	D	E	F	G	H	I
1	Problem	i	N	PMT	PV	FV	Solve for	Answer	Formula
2	5-8	10%	10	18000		100,000	PV	$149,157	=-PV(B2,C2,D2,F2)

PV = $149,157.

5-9

Your company has been presented with an opportunity to invest in the following project. The facts on the project are:

Investment required	$90,000,000
Salvage value after 10 years	0
Gross income	20,000,000
Annual operating costs:	
Labor	2,500,000
Materials, licenses, insurance, etc.*	1,000,000
Fuel and other costs	1,500,000
Maintenance costs	500,000

*Beginning-of-period cash flow

The project will operate for 10 years. If management expects 8% on its investments before taxes, would you recommend this project?

Solution

$$PW = -90,000,000 + [20,000,000 - 4,500,000 - 1,000,000(F/P, 8\%, 1)](P/A, 8\%, 10)$$
$$= \$6,758,200$$

Accept the project because it has a positive NPW.

5-10

Sarah Bishop, having become a very successful engineer, wishes to start an endowment at UTM that will provide scholarships of $10,000 per year to four engineering students beginning in Year 6 and continuing indefinitely. Determine the amount Sarah must donate now if the university earns 10% per year on endowment funds.

Solution

$10,000 \times 4 = 40,000$

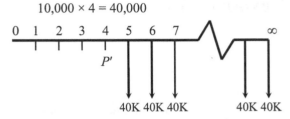

Amount needed at end of Year 4 $P' = 40,000(P/A, 10\%, \infty)$

$= 40,000(1/0.1)$

$= 400,000$

Amount needed today $P = 400,000(P/F, 10\%, 4)$

$= \$273,200$

5-11

What is the maximum amount you should be willing to pay for the following piece of equipment? First-year income is anticipated to be $18,000 and decrease by $1500 each year thereafter. First-year costs are anticipated to be $5000 and increase by $500 each year thereafter. The salvage value is estimated to be 8.5% of the first cost. Use an interest rate of 4% and useful life of 5 years.

Solution

$0 = -FC + [18,000(P/A, 4\%, 5) - 1500(P/G, 4\%, 5)]$

$- [5000 \ (P/A, 4\%, 5) + 500(P/G, 4\%, 5)] + .085 \ FC(P/F, 4\%, 5)$

$FC = \$43,830$

5-12

Find the present worth of the following cash flow diagram if $i = 8\%$.

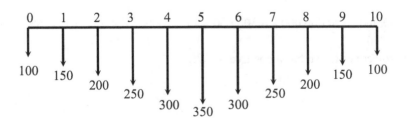

Solution

$$P = 100 + 150(P/A, 8\%, 5) + 50(P/G, 8\%, 5)$$
$$+ [300(P/A, 8\%, 5) - 50(P/G, 8\%, 5)](P/F, 8\%, 5) = \$1631.97$$

5-13

A couple wants to begin saving money for their daughter's education. $16,000 will be needed on the child's 18th birthday, $18,000 on the 19th birthday, $20,000 on the 20th birthday, and $22,000 on the 21st birthday. Assume 5% interest with annual compounding. The couple is considering two methods of accumulating the money.

    a. How much money would have to be deposited into the account on the child's first birthday to accumulate enough money to cover the education expenses? (*Note*: A child's "first birthday" is celebrated 1 year after the child is born.)

    b. What uniform annual amount would the couple have to deposit each year on the child's first through seventeenth birthdays to accumulate enough money to cover the education expenses?

Solution

    a.

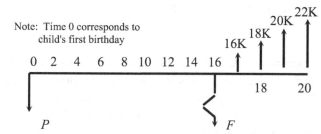

Let F = the number of dollars needed at the beginning of Year 16
$$= 16,000(P/A, 5\%, 4) + 2000(P/G, 5\%, 4)$$
$$= \$66,942$$

The amount needed today: $P = 66,942(P/F, 5\%, 16) = \$30,666.13$

    b.

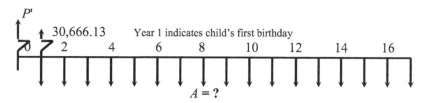

$P' = 30,666.13(P/F, 5\%, 1) = \$29,206.42$
$A = 19,206.42(A/P, 5\%, 17) = \$2,590.61$

5-14

Assume that you borrowed $50,000 at an interest rate of 1% per month, to be repaid in uniform monthly payments for 30 years. How much of the 163rd payment would be interest, and how much would be principal?

Solution

The interest paid on a loan at time t is determined by multiplying the effective interest rate times the outstanding principal just after the preceding payment at time $t - 1$.

To find the interest paid at time $t = 163$ (call it I_{163}), first find the outstanding principal at time $t = 162$ (call it P_{162}). This can be done by computing the future worth at time $t = 162$ of the amount borrowed, minus the future worth of 162 payments.

Compute the present worth, at time 162, of the 198 payments remaining.
The uniform monthly payments are 50,000(A/P, 1%, 360) = $514.31. Therefore,
$P_{162} = 514.31(P/A, 1\%, 198) = \$44,259.78$
The interest $I_{163} = 0.01(44,259.78) = \$442.60$
The principal in the payment is $514.31 - 442.60 = \$71.71$.

5-15

A town is seeking a new tourist attraction, and the town council has voted to budget $500,000 for the project. A survey shows that an interesting cave can be enlarged and developed for a contract price of $400,000. The proposed attraction is expected to have an infinite life. The estimated annual expenses of operation total $50,000. The price per ticket is to be based on an average of 12,000 visitors per year. If money is worth 8%, what should be the price of each ticket?

Solution

If the $100,000 cash, left over after developing the cave, is invested at 8%, it will yield a perpetual annual income of $8000. This $8000 can be put toward the expenses ($50,000 a year). The balance of the expenses can be raised through ticket sales, making the price per ticket:
Ticket price = $42,000/12,000 tickets = $3.50

Alternate solution:
$PW_{COST} = PW_{BENEFIT}$
$400,000 + (50,000)/0.08 = 500,000 + T/0.08$
$400,000 + 625,000 = 500,000 + T/0.08$
$T = 525,000(0.08) = \$42,000$
Ticket price = $42,000/12,000 tickets = $3.50.

5-16

Two alternatives are being considered for recovering aluminum from garbage. Alternative 1 has a capital cost of $100,000 and a first-year maintenance cost of $12,000, with maintenance increasing by $500 per year for each year after the first.

Alternative 2 has a capital cost of $140,000 and a first-year maintenance cost of $4000, with maintenance increasing by $1000 per year after the first.

Revenues from the sale of aluminum are $20,000 in the first year, increasing $2000 per year for each year after the first. The life of both alternatives is 10 years. There is no salvage value. The before-tax MARR is 4%. Use present worth analysis to determine which alternative is preferred.

Solution

Alternative 1:
NPW $= -100,000 + 8000(P/A, 4\%, 10) + 1500(P/G, 4\%, 10) = \$15,709.50$
Alternative 2:
NPW $= -140,000 + 16,000(P/A, 4\%, 10) + 1000(P/G, 4\%, 10) = \$23,657.00$
Choose Alternative 2 $\rightarrow$ maximum NPW.

5-17

As a temporary measure, a brewing company is deciding between two used filling machines: the Kram and the Zanni.
a. The Kram filler has an initial cost of $85,000; the estimated annual maintenance is $8000.
b. The Zanni filler has a purchase price of $42,000, with annual maintenance costs of $8000.

The Kram filler has a higher efficiency than the Zanni, and it is expected that savings would amount to $4000 per year if the Kram filler were installed. The filling machine will not be needed after 5 years, and at that time, the salvage value for the Kram filler would be $25,000, while the Zanni would have little or no value. Assuming a MARR of 10%, which filling machine should be purchased?

Solution

There is a fixed output; therefore, minimize costs.
Kram:

NPW $= 25,000(P/F, 10\%, 5) - 85,000 - 4000(P/A, 10\%, 5)$
$= -84,641.5$ (or a PW$_{\text{Cost}}$ of $84,641.50)

Zanni:

NPW $= -42,000 - 8000(P/A, 10\%, 5)$
$= -72,328$ (or a PW$_{\text{Cost}}$ of $72,328)

Therefore choose the Zanni filler.

5-18

McClain, Edwards, Shiver, and Smith (MESS) LLC is considering the purchase of new automated cleaning equipment. The industrial engineer for the company has been asked to calculate the present worth of the two best alternatives based on the following data.

	Mess Away	Quick Clean
First cost	$65,000	$78,000
Annual savings	20,000	24,000
Annual operating costs	4,000	2,750
Scheduled maintenance	$1500 at the end of 3rd year	$3000 at the end of 3rd year
Annual insurance*	2,000	2,200
Salvage value	10% of first cost	12.5% of first cost
Useful life	5 years	5 years

　* Assume beginning-of-period payments.

Determine which equipment should be purchased, given an interest rate of 8%.

Solution

Mess Away

Year		
0	First cost	(65,000)
1–5	Annual net savings: 16,000(P/A, 8%, 5)	63,888
0–4	Annual insurance: 2000 + 2000(P/A, 8%, 4)	(8,624)
3	Scheduled maintenance: 1500(P/F, 8%, 3)	(1,191)
5	Salvage value: 6500(P/F, 8%, 5)	4,424
		$(6,503)

Quick Clean

Year		
0	First cost	(78,000)
1–5	Annual net savings: 21,250(P/A, 8%, 5)	4,851
0–4	Annual insurance: 2200 + 2200(P/A, 8%, 4)	(9,486)
3	Scheduled maintenance: 3000(P/F, 8%, 3)	(2,381)
5	Salvage value: 10,000(P/F, 8%, 5)	6,806
		$ 1,790

Choose Quick Clean.

5-19

Helbing Construction has just been awarded a contract that will require the company to either purchase or lease a new track hoe. The equipment will be needed for the entire length of the 4-year contract. The equipment can be leased for $85,000 per year. All operating costs must be paid by Helbing. These are estimated to be $56,000 per year. All other expenses will be paid by the leasing company. If the track hoe is purchased, the first cost will be $485,000 and the anticipated salvage value is $365,000 at the end of the 4 years. The annual operating costs are expected to be $75,000 with an additional maintenance costing $5000 occurring at the end of Year 2. If Helbing's MARR is 5%, should the company lease or purchase the track hoe?

Solution

Lease

Year		
$0-3$	Lease payment $85,000 + 85,000(P/A, 5\%, 3)$	$(316,455)$
$1-4$	Operating costs $56,000(P/A, 5\%, 4)$	$\underline{(198,576)}$
		$\$(515,031)$

Purchase

Year		
0	First cost	$(485,000)$
$1-4$	Operating costs $75,000(P/A, 5\%, 4)$	$(265,950)$
2	Maintenance $5000(P/F, 5\%, 2)$	$(4,535)$
4	Salvage value $365,000(P/F, 5\%, 4)$	$\underline{300,286}$
		$\$(455,199)$

Purchase the equipment.

5-20

Gullett Glue (GG) must replace a machine used to fill glue tubes. The relevant data concerning the machines under consideration are presented in the table below. If the MARR for GG is 5%, determine which machine should be purchased.

	Fill-Rite	Best-Fill
First cost	$72,000	$68,000
Annual costs	$6,000 the first year increasing by $750 each year thereafter	$7,800 per year
Overhaul	$7,525 at the end of Year 4	$8,000 at the end of Year 4
Salvage value	7.5% of first cost	$5,500
Useful life	8 years	8 years

Solution

$PW_{F-R} = -72,000 - [6000(P/A, 5\%, 8) + 750(P/G, 5\%, 8)] - 7525(P/F, 5\%, 4)$
$$+ .075(72,000)(P/A, 5\%, 8)$$
$$= -\$129,042$$
$PW_{B-F} = -68,000 - 7800(P/A, 5\%, 8) - 8000(P/F, 5\%, 4) + 5500(P/A, 5\%, 8)$
$$= -\$121,271$$
Choose Best-Fill.

5-21

Two alternatives are under consideration by XYZ Inc. Alternative A has a NPV of $1243 and a life of 4 years. Alternative B has a NPV of $2196 and a life of 8 years. Which alternative should XYZ select if the MARR is 5%?

Solution

$NPW_{A8} = 1243 + 1243(P/F, 5\%, 4) = \$2265.62$
Choose Alternative B $\rightarrow$ maximum NPW.

5-22

A project has a first cost of $10,000, net annual benefits of $2000, and a salvage value of $3000 at the end of its 10-year useful life. The project will be replaced identically at the end of 10 years, and again at the end of 20 years. What is the present worth of the entire 30 years of service, given an interest rate of 10%?

Solution

PW of 10 years $= -10,000 + 2000(P/A, 10\%, 10) + 3000(P/F, 10\%, 10) = \$3445.76$
PW of 30 years $= 3445.76 + 3445.76\ (P/F, 10\%, 10) + 3445.76\ (P/F, 10\%, 20) = \$5286.45$

5-23

Using a 10% interest rate, determine which alternative, if any, should be selected.

Alternative	_A_	_B_
First cost	$5300	$10,700
Uniform annual benefit	1800	2,100
Useful life	4 years	8 years

Solution

Alternative A:
$NPW = -5300 + 1800(P/A, 10\%, 8) - 5300(P/F, 10\%, 4) = \$683.10$
Alternative B:
$NPW = -10,700 + 2100(P/A, 10\%, 8) = \$503.50$
Select Alternative A.

5-24

The lining of a chemical tank in a certain manufacturing operation is replaced every 5 years at a cost of $5000. A new type of lining is now available that would last 10 years, but it costs $9000. The manufacturer's tank needs a new lining now, and the company intends to use the tank for 40 years, replacing linings when necessary. Compute the present worth of costs of 40 years of service for the 5-year and 10-year linings if $i = 10\%$.

Solution

    PW 5-year lining:

        PW = [5000(A/P, 10%, 5)](P/A, 10%, 40) = $12,898.50

    PW 10-year lining:

        PW = [9000(A/P, 10%, 10)](P/A, 10%, 40) = $14,319.39

5-25

Be-low Mining Inc., is trying to decide whether it should purchase or lease new earthmoving equipment. If purchased, the equipment will cost $175,000 and will be used 6 years, at which time it can be sold for $72,000. At Year 3, an overhaul costing $20,000 must be performed. The equipment can be leased for $30,000 per year. Be-low will not be responsible for the midlife overhaul if the equipment is leased. If the equipment is purchased, it will be leased to other mining companies when possible; this is expected to yield revenues of $15,000 per year. The annual operating cost regardless of the decision will be approximately equal. What would you recommend if the MARR is 6%?

Solution

    Lease: (Recall that lease payments are beginning-of-period cash flows.)

    $P = -30,000 - 30,000(P/A, 6\%, 5) = -\$156,360$

    Buy:

    $P = -175,000 + 72,000(P/F, 6\%, 6) - 20,000(P/F, 6\%, 3) + 15,000(P/A, 6\%, 6)$

       $= -\$67,277$

    Recommend the buy option.

5-26

Two technologies are currently available for the manufacture of an important and expensive food and drug additive.

Laboratory A is willing to release the exclusive right to manufacture the additive in this country for $50,000 payable immediately, and a $40,000 payment each year for the next 10 years. The production costs are $1.23 per unit of product.

Laboratory B is also willing to release similar manufacturing rights, with the following schedule of payments:

> on the closing of the contract, $10,000
> from Years 1 to 5, at the end of each year, a payment of $25,000 each
> from Years 6 to 10, also at the end of each year, a payment of $20,000

The production costs are $1.37 per unit of product.

Neither lab is to receive any money after 10 years for this contract. It is anticipated there will be an annual production of 100,000 items for the next 10 years. On the basis of analyses and trials, the products of A and B are practically identical in quality. Assuming a MARR of 12%, which lab should be chosen?

Solution

Laboratory A: The annual production cost $= 1.23 \times 100K = \$123K$

$PW_{Cost} = 50,000 + [40,000 + 123,000](P/A, 12\%, 10) = \$970,950$

Laboratory B: The annual production cost $= 1.37 \times 100K = \$137K$

$PW_{Cost} = 10,000 + [25,000 + 137,000](P/A, 12\%, 5)$
$\qquad\qquad + [20,000 + 137,000](P/A, 12\%, 5)(P/F, 12\%, 5) = \$915,150$

Therefore, choose Laboratory B.

5-27

An engineering analysis by net present worth (NPW) is to be made for the purchase of two devices, A and B. If an 8% interest rate is used, recommend the device to be purchased.

	Cost	Uniform Annual Benefit	Salvage	Useful Life
Device A	$600	$100	$250	5 years
Device B	700	100	180	10 years

Solution

Device A:

$NPW = 100(P/A, 8\%, 10) + 250(P/F, 8\%, 10) - 600 - [600 - 250](P/F, 8\%, 5) = -\$51.41$

Device B:

$NPW = 100(P/A, 8\%, 10) + 180(P/F, 8\%, 10) - 700 = \$54.38$

Select Device B.

5-28

An engineer is considering buying a life insurance policy for his family. He currently owes $50,500, and he would like his family to have an annual available income of $55,000 indefinitely upon receipt of the policy proceeds. If the engineer believes that money from the policy can be invested in an account paying 4% annual interest, how much life insurance should he buy?

Solution
> 4% interest $n = \infty$
> $P = A/i = 55{,}000/0.04 = 1{,}375{,}000$
> Total life insurance $= 50{,}500 + 1{,}375{,}000 = \$1{,}425{,}500$

5-29

A successful engineer wishes to establish a scholarship at her alma mater that will pay two industrial engineering students $4000 per year. The university earns 5% on endowment accounts. Determine the amount that must be deposited today if the scholarships will be awarded 1 year from today.

Solution
$$P = 8000(P/A, 5\%, \infty) = \$160{,}000$$

5-30

A bridge is being considered at a cost of $220M. The annual maintenance costs are estimated to be $150K. A major renovation costing $50M is required every 25 years. What is the capitalized cost of the bridge at 5% interest?

Solution
> Capitalized cost $= 220M + 0.15M(P/A, 5\%, \infty) + 50M(A/F, 5\%, 25)(P/A, 5\%, \infty)$
> $\qquad\qquad = 220M + 0.15M(1/0.05) + 50M(0.021)(1/0.05) = \$244M$

5-31

A resident will give money to his town to purchase a statue honoring the town founders and will pay to maintain the work at a cost of $500 per year forever. If an interest rate of 5% is used, and the resident gives a total of $25,000; how much can be paid for the statue?

Solution
> Capitalized cost $= 25{,}000 = P + 500(P/A, 5\%, \infty)$
> $P = 25{,}000 - 500(1/0.05) = \$15{,}000$

5-32

A minor league baseball stadium is built at a cost of $12,000,000. The annual maintenance costs are expected to cycle every 6 years with a cost of $18,000 the first year and $2000 more each year thereafter until the cost is $28,000 at the end of Year 6. The maintenance costs cycle will repeat starting with $18,000 in Year 7. Every 6 years an expenditure of $1,500,000 must be made to renovate and update the stadium. Determine the capitalized cost at 4%.

Solution

Capitalized cost = 12,000,000 + [18,000 + 2000(A/G, 4%, 6)](P/A, 4%, ∞)
+ 1,500,000(A/F, 4%, 6) (P/A, 4%, ∞)
= $18,224,300

5-33

A tunnel transporting water through a mountain range initially cost $1,000,000 and has expected maintenance costs that will occur in a 6-year cycle as shown.

End of Year:	1	2	3	4	5	6
Maintenance:	$35,000	$35,000	$35,000	$45,000	$45,000	$60,000

Determine the capitalized cost at 8% interest.

Solution

Capitalized cost = PW of cost for an infinite time.
First compute the equivalent annual cost (EAC) of the maintenance.
EAC = 35,000 + [10,000(F/A, 8%, 3) + 15,000](A/F, 8%, 6) = $41,468.80
For $n = ∞$, $P = A/i$
Capitalized cost = 1,000,000 + (41,468.80/0.08) = $1,518,360

5-34

A new runway at Chester International Airport was recently built at a cost of $16,000,000. Maintenance and upkeep costs will be $450,000 per year. The runway lighting will require replacement at a projected cost of $2,000,000 every 4 years. Every 8 years the runway will require resurfacing at a projected cost of $6,000,000. Determine the capitalized cost of the runway at an interest rate of 5%.

Solution

Capitalized cost = 16,000,000 + 450,000(P/A, 5%, ∞) + 2,000,000(A/F, 5%, 4)(P/A, 5%, ∞)
+ 6,000,000(A/F, 5%, 8)(P/A, 5%, ∞)
= $1,164,912

5-35

Cheap Motors Manufacturing must replace one of its tow motors. The NPW of Alternative *A* is – $5876; for Alternative *B* it is –$7547; and for Alternative *C*, –$3409. Alternatives *A* and *B* are expected to last for 12 years, and Alternative *C* is expected to last for 6 years. If Cheap's MARR is 4%, which alternative should be chosen?

Solution

A 12-year analysis period is necessary.

$NPW_{A12} = -\$5876$

$NPW_{B12} = -\$7547$

$NPW_{C12} = -3409 + 3409(P/F, 4\%, 6) = -\$6103$

If an alternative must be chosen, minimize the PW of costs; choose Alternative A.

5-36

A company decides that it <u>must</u> provide repair service for the equipment it sells. Based on the following, which alternative for providing repair service should be selected?

Alternative	NPW
A	–$9241
B	–6657
C	–8945

Solution

None of the alternatives look desirable; but since one must be chosen (the do-nothing alternative is not available), choose the one that maximizes NPW (in this case, minimizes net present costs). The best of the alternatives is *B*.

5-37

A farmer must borrow $20,000 to purchase a used tractor. The bank has offered the following choice of payment plans, each determined by using an interest rate of 8%. If the farmer's MARR is 15%, which plan should he choose?

Plan *A*: $5010 per year for 5 years

Plan *B*: $2956 per year for 4 years plus $15,000 at end of 5 years

Plan *C*:   Nothing for 2 years, then $9048 per year for 3 years

Solution

$PW_{Cost-A} = 5010(P/A, 15\%, 5) = \$16,794$

$PW_{Cost-B} = 2956(P/A, 15\%, 4) + 15,000(P/F, 15\%, 5) = \$15,897$

$PW_{Cost-C} = 9048(P/A, 15\%, 3)(P/F, 15\%, 2) = \$15,618$

Plan *C* is lowest-cost plan.

5-38

A firm is considering the purchase of a new machine to increase the output of an existing production process. Of all the machines considered, management has narrowed the field to the machines with the following cash flows.

Machine	Initial Investment	Annual Operating Income
1	$ 50,000	$22,815
2	60,000	25,995
3	75,000	32,116
4	80,000	34,371
3	100,000	42,485

If each of these machines provides the same service for 3 years and the minimum attractive rate of return is 6%, which machine should be selected?

Solution

Maximize the PW.

Machine	Initial Investment	Operating Income (P/A, 6%, 3)	NPW
1	$ –60,000	+ 22,565(2.673)	= $316
2	–65,000	+ 24,445(2.673)	= $342
3	–75,000	+ 28,125(2.673)	= $178
4	–80,000	+ 30,010(2.673)	= $217
5	–100,000	+ 37,450(2.673)	= $104

Select Machine 2.

5-39

The city council wants the municipal engineer to evaluate three alternatives for supplementing the city water supply. The first alternative is to continue deep-well pumping at an annual cost of $10,500. The second alternative is to install an 18-inch pipeline from a surface reservoir. First cost is $25,000 and annual pumping cost is $7000.

The third alternative is to install a 24-inch pipeline from the reservoir at a first cost of $34,000 and annual pumping cost of $5000. The life of each alternative is 20 years. For the second and third alternatives, salvage value is 10% of first cost. With interest at 8%, which alternative should the engineer recommend? Use present worth analysis.

Solution

Fixed output, therefore minimize cost.

Year	Deep Well	18-in. Pipeline	24-in. Pipeline
0		−25,000	−34,000
1–20	−10,500	−7,000	−5,000
20		+2,500	+3,400

Deep Well: PWC = −10,500(*P/A*, 8%, 20) = −$103,089
18-in. Pipeline: PW of cost = −25,000 − 7000(*P/A*, 8%, 20) + 2500(*P/F*, 8%, 20) = −$93,190
24-in. Pipeline: PW of cost = −34,000 − 5000(*P/A*, 8%, 20) + 3400(*P/F*, 8%, 20) = −$82,361
Choose the 24-inch pipeline.

5-40

The following data are associated with three grape-crushing machines under consideration by Rabbit Ridge Wineries LLC.

	Smart Crush	Super Crush	Savage Crush
First cost	$52,000	$63,000	$105,000
O&M costs	15,000	9,000	12,000
Annual benefits	38,000	31,000	37,000
Salvage value	13,000	19,000	22,000
Useful life	4 years	6 years	12 years

If Rabbit Ridge uses a MARR of 12%, which alternative, if any, should be chosen?

Solution

A 12-year analysis period is necessary.
Smart Crush:
$NPW_4 = -52,000 + 23,000(P/A, 12\%, 4) + 13,000(P/F, 12\%, 4) = \$26,113$
$NPW_{12} = 26,113 + 26,113(P/F, 12\%, 4) + 26,113(P/F, 12\%, 8) = \$53,255$
Super Crush:
$NPW_6 = -63,000 + 22,000(P/A, 12\%, 6) + 19,000(P/F, 12\%, 6) = \$37,067$
$NPW_{12} = 37,067 + 37,067(P/F, 12\%,6) = \$55,845$
Savage Crush:
$NPW_{12} = -105,000 + 25,000(P/A, 12\%, 12) + 22,000(P/F, 12\%, 12) = \$54,497$

To maximize NPW, choose Super Crush.

5-41

A used car dealer states that if you put $2000 down on a particular car, your monthly payments will be $199.08 for 4 years at an interest rate of 9%. What is the cost of the car to you?

Solution

A = $199.08 per period $i = 9\%/12 = \frac{3}{4}\%$ $n = 12 \times 4 = 48$

P = $2000 + 190.08(P/A, \frac{3}{4}\%, 48) = \$10,000$

5-42

Mary Ann requires approximately 30 pounds of bananas each month, January through June, and 35 pounds of bananas each month, July through December, to make banana cream pies for her friends. Bananas can be bought at a local market for 40 cents/lb. If Mary Ann's cost of money is 3%, approximately how much should she set aside at the beginning of each year to pay for the bananas?

Solution

Cost of bananas   January–June $30 \times 0.40 = \$12$

                    July–December $35 \times 0.40 = \$14$

$i = 3\%/12 = \frac{1}{4}\%$

$P = 12(P/A, \frac{1}{4}\%, 6) + 14(P/A, \frac{1}{4}\%, 6)(P/F, \frac{1}{4}\%, 6) = \$153.41$

5-43

Three purchase plans are available for a new car.

Plan A: $5000 cash immediately

Plan B: $1500 down and 36 monthly payments of $97.75

Plan C: $1000 down and 48 monthly payments of $96.50

If a customer expects to keep the car for 5 years, and her cost of money is 6%, which payment plan should she choose?

Solution

Note that in all cases the car is kept for 5 years; that is the common analysis period.

$i = 6\%/12 = \frac{1}{2}\%$

$PW_A = \$5000$

$PW_B = 1500 + 97.75(P/A, \frac{1}{2}\%, 36) = \$4713.14$

$PW_C = 1000 + 96.50(P/A, \frac{1}{2}\%, 48) = \$5108.97$

Therefore, Plan B is the best plan.

5-44

A firm has the following monthly costs. Find the present worth of the costs if the annual interest rate is 15%.

Solution

	A	B	C	D
1	15%	Annual Interest		
2	1.3%	Monthly Interest		
3				
4	N	Month	Costs	PW
5	1	January	$15,000	$14,815
6	2	Febuary	$25,000	$24,387
7	3	March	$205,000	$197,501
8	4	April	$290,000	$275,942
9	5	May	$280,000	$263,138
10	6	June	$140,000	$129,944
11	7	July	$80,000	$73,337
12	8	August	$50,000	$45,270
13				
14			Total PW =	$1,024,333

The present worth is $1,024,333.

5-45

Parker Designs has a profitable project, and the customer has agreed to payments as follows. Use the XNPV function to find the present worth of the cash flows as of January 1, 2019, if the interest rate is 8.0%.

Date	Income
1/1/2019	$75,000
5/6/2019	75,000
7/8/2019	75,000
9/2/2019	75,000
12/2/2019	100,000

Solution

◢	A	B	C	D
1	Interest Rate	8.0%		
2				
3	Date	Cash Flow		
4	01/01/2019	$75,000		
5	05/06/2019	75,000		
6	07/08/2019	75,000		
7	09/02/2019	75,000		
8	12/02/2019	100,000		
9				
10	$ 384,553	=XNPV(B1,B4:B8,A4:A8)		

The present worth is $384,553.

5-46

A retail sunscreen's manufacturer has income for one season as shown. Use XNPV to find the present worth as of March 1, 2019 if the MARR is 15%.

Date	Income
3/1/2019	−$15.0 million
4/1/2019	6.1 million
5/6/2019	1.3 million
6/3/2019	3.5 million
7/1/2019	4.6 million
8/5/2019	2.5 million
9/2/2019	0.9 million

Solution

▲	A	B	C	D
1	Interest Rate	15.0%		
2				
3	Date	Cash Flow (in Millions)		
4	03/01/2019	-$15.0		
5	04/01/2019	6.1		
6	05/06/2019	1.3		
7	06/03/2019	3.5		
8	07/01/2019	4.6		
9	08/05/2019	2.5		
10	09/02/2019	0.9		
11				
12	$	3.25 =XNPV(B1,B4:B10,A4:A10)		

The present worth is $3.25 million.

5-47

If the current market interest rate on bonds of a certain type is 3%, compounded semiannually, what should be the market price of a 5% bond having a $1000 face value? The bond will mature (pay its face value) 6½ years from today, and the next interest payment to the bondholder will be due in 6 months.

Solution

Semiannual interest payment = 0.025(1000) = $25

$PV = \$25(P/A, 1.5\%, 13) + \$1000(P/F, 1.5\%, 13) = \$1117$

Using a spreadsheet,

▲	A	B	C	D	E	F	G	H	I
1	Problem	i	N	PMT	PV	FV	Solve for	Answer	Formula
2	5-47	1.5%	13	25		1,000	PV	$1,117	=-PV(B2,C2,D2,F2)

5-48

A bond issued by Golden Key casinos has a face value of $100,000 and a face rate of 3.75% payable semiannually. If the bond matures 6 years from the date of purchase and an investor requires a 5% return, what is the maximum the investor should pay for the bond?

Solution

Semi-annual payment = $(0.0375/2)(100,000) = \$1875$

$PW_{BOND} = -\text{Cost} + 1875(P/A, 2\tfrac{1}{2}\%, 12) + 100,000(P/F, 2\tfrac{1}{2}\%, 12)$

In order to determine the maximum price, set the $PW_{BOND} = 0$

$0 = -\text{Cost} + 1875(P/A, 2\tfrac{1}{2}\%, 12) + 100,000(P/F, 2\tfrac{1}{2}\%, 12)$

$\text{Cost} = \$93,589$

Chapter 6
Annual Cash Flow Analysis

6-1

Deere Construction just purchased a new track hoe attachment costing $12,500. The CFO expects the implement will be used for 5 years, at which time its salvage value is estimated to be $4000. Maintenance costs are estimated at $0 the first year, increasing by $100 each year thereafter. If a 12% interest rate is used, what is the equivalent uniform annual cost of the implement?

Solution

$$EUAC = 12,500(A/P, 12\%, 5) - 4000(A/F, 12\%, 5) + 100(A/G, 12\%, 5) = \$3015.40$$

6-2

The survey firm of Myers, Anderson, and Pope (MAP) LLP is considering the purchase of new GPS equipment. Data concerning the alternative under consideration are as follows.

First cost	$28,000
Annual income	7,000
Annual costs	2,500
Recalibration at end of Year 4	4,000
Salvage value	2,800

If the equipment has a life of 8 years and MAP's minimum attractive rate of return (MARR) is 5%, what is the annual worth of the equipment?

Solution

$$EUAC = 28,000(A/P, 5\%, 8) - 4500 - 4000(P/F, 5\%, 4)(A/P, 5\%, 8) - 2800(A/F, 5\%, 8) = -\$47.63$$

6-3

Ronald McDonald decides to install a fuel storage system for his farm that will save him an estimated 6.5 cents/gallon on his fuel cost. He uses an estimated 20,000 gallons/year on his farm. Initial cost of the system is $10,000, and the annual maintenance the first year is $25, increasing by $25 each year thereafter. After a period of 10 years the estimated salvage is $3000. If money is worth 12%, is the new system a wise investment?

Solution

$$\text{EUAC} = 10,000(A/P, 12\%, 10) + 25 + 25(A/G, 12\%, 10)$$
$$= \$1884.63$$
$$\text{EUAB} = 20,000(0.065) + 3000(A/F, 12\%, 10) = \$1471.00$$
$$\text{EUAW} = -\$413.63 \therefore \text{ not a wise investment}$$

6-4

The incomes for a business for 5 years are as follows: $8250, $12,600, $9,750, $11,400, and $14,500. If the value of money is 8%, what is the equivalent uniform annual benefit for the 5-year period?

Solution

$$\text{PW} = 8250(P/F, 8\%, 1) + 12,600(P/F, 8\%, 2) + 9750(P/F, 8\%, 3)$$
$$+ 11,400(P/F, 8\%, 4) + 14,500(P/F, 8\%, 5) = \$44,427.91$$
$$\text{EUAB} = 44,427.91(A/P, 8\%, 5) = \$11,129.19$$

6-5

For the cash flow diagram shown, write the equation that properly calculates the uniform equivalent.

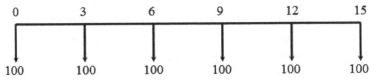

Solution

$$A = 100(A/P, i, 15) + 100(A/F, i, 3)$$

6-6

A project has a first cost of $75,000, operating and maintenance costs of $10,000 during each year of its 8-year life, and a $15,000 salvage value. If the interest rate is 12%, what is its equivalent uniform annual cost (EUAC)?

Solution

$$\text{EUAC} = 75,000(A/P, 12\%, 8) + 10,000 - 15,000(A/F, 12\%, 8) = \$23,878$$

6-7

A foundation supports an annual campus seminar by using the earnings of a $50,000 gift. It is felt that 10% interest will be realized for 10 years but that plans should be made to anticipate an interest rate of 6% after that time. What uniform annual payment may be established from the beginning, to fund the seminar at the same level into infinity?

Solution

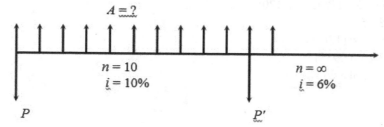

Assume first seminar occurs at time of deposit.

$$P' = A/i = A/0.06$$
$$P = A + A(P/A, 10\%, 10) + P'(P/F, 10\%, 10)$$
$$50,000 = A + 6.145A + (A/.06)0.3855$$
$$13.57A = 50,000$$
$$A = \$3,684.60$$

6-8

A project requires an initial investment of $10,000 and returns benefits of $3000 at the end of every fifth year thereafter. If the minimum attractive rate of return (MARR) is 5%, calculate the equivalent uniform annual worth.

Solution

Year:	0	5	10	15	20	25 ...	... ∞
Cash Flow ($):	−10,000	6000	6000	6000	6000	6000	6000

$$\text{EUAW} = 6000(A/F, 5\%, 5) - 10,000(A/P, 5\%, \infty) = \$586$$

6-9

At an interest rate of 10% per year, determine the perpetual equivalent annual cost of: $70,000 now, $100,000 at the end of Year 6, and $10,000 per year from the end of Year 10 through infinity.

Solution

$$P = 70,000 + 100,000(P/F, 10\%, 6) + 10,000(P/A, 10\%, \infty)(P/F, 10\%, 10) = \$165,110$$
$$A = 165,110(A/P, 10\%, \infty) = \$16,511$$

6-10

A recent engineering graduate makes a donation of $20,000 now and will pay $375 per month for 10 years to endow a scholarship. If interest is a nominal 9%, what annual amount can be awarded? Assume that the first scholarship will be bestowed at the end of the first year.

Solution

$P = 20{,}000 + 375(P/A, \frac{3}{4}\%, 120)$

    $= 49{,}603.25$

$A = Pi$ where $i = (1.0075)^{12} - 1$

    $= 49{,}603.25(0.0938)$

    $= \$5952$ scholarship

6-11

Given

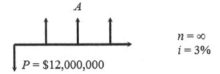

$n = \infty$
$i = 3\%$

$P = \$12{,}000{,}000$

Find: A

Solution

$A = Pi$

    $= 12{,}000{,}000(0.03)$

    $= \$360{,}000$

6-12

The first cost of a fairly large flood control dam is expected to be $5 million. The maintenance cost will be $60,000 per year, and a $100,000 outlay will be required every 5 years. At interest of 10%, find the EUAC of the dam project.

Solution

$\text{EUAC} = 5{,}000{,}000(A/P, 10\%, \infty) + 60{,}000 + 100{,}000(A/F, 10\%, 5) = \$576{,}380$

6-13

Determine the equivalent annual worth of a 5-year lease with annual payments of $5000 at 5%.

Solution

Recall lease payments are beginning-of-period and annual worth is end-of-period.

The payment is converted to an equivalent end-of-year value by using $(F/P, i\%, 1)$.

$A = 5000(F/P, 5\%, 1) = 500{,}000(1.05) = \$42{,}000$

6-14

Twenty-five thousand dollars is deposited in a bank trust account that pays 3% interest, compounded semiannually. Equal annual withdrawals are to be made from the account, beginning one year from now and continuing forever. Calculate the maximum amount of W, the annual withdrawal.

Solution

$i = 3\%/2 = 1\frac{1}{2}\%$

$A = Pi = 25,000(0.015) = 375$ per semiannual period

$W = 375(F/A, 1\frac{1}{2}\%, 2) = 375(2.015) = \$755.62$

6-15

A tractor costs \$32,500 and will be used for 5 years, at which time its estimated salvage value will be \$14,000. Maintenance costs are estimated to be \$500 for the first year, increasing by \$150 each year thereafter. If $i = 6\%$, what is the equivalent uniform annual cost (EUAC) for the tractor?

Solution

EUAC $= 32,500(A/P, 6\%, 5) + 500 + 150(A/G, 6\%, 5) - 14,000(A/F, 6\%, 5)$
$= \$6014.50$

6-16

If in the last week of February 2016 Ellen won \$250,000 and invested it by March 1, 2016, in a "sure thing" that paid 8% interest, compounded annually, what uniform annual amount can she withdraw on the first of March for 15 years starting in 2022?

Solution

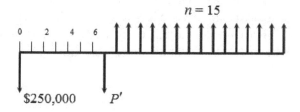

$P' = 250,000(F/P, 8\%, 7) = \$428,500$

$A = 250,000(F/P, 8\%, 7)(A/P, 8\%, 15) = \$50,048.80$

6-17

A machine having a first cost of $20,000 is expected to save $1500 in the first year of operation, and the savings should increase by $200 every year until (and including) the 9th year; thereafter, the savings will decrease by $150 until (and including) the 16th year. Using equivalent uniform annual worth, is this machine economical? Assume a MARR of 10%.

Solution

There are a number of possible solutions. Here's one:

$$\text{EUAW} = -20,000(A/P, 10\%, 16) + [1500(P/A, 10\%, 9) + 200(P/G, 10\%, 9)](A/P, 10\%, 16)$$
$$+ [2950(P/A, 10\%, 7) - 150(P/G, 10\%, 7)](P/F, 10\%, 9)(A/P, 10\%, 16)$$
$$= -\$280.94; \text{ the machine is not economical}$$

6-18

Calculate the equivalent uniform annual cost of the following schedule of payments.

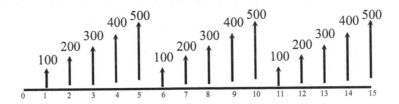

Solution

Since payments repeat every 5 years, analyze for 5 years only.

$$A = 100 + 100(A/G, 8\%, 5) = \$284.60$$

6-19

The initial cost of a van is $22,800; its salvage value after 5 years will be $8500. Maintenance is estimated to be a uniform gradient amount of $225 per year (with no maintenance costs the first year), and the operation cost is estimated to be 36 cents/mile for 400 miles/month. If money is worth 6%, what is the approximate equivalent uniform annual cost (EUAC) for the van, expressed as a monthly cost?

Solution

$$\text{EUAC} = 22,800(A/P, 6\%, 5) + 225(A/G, 6\%, 5) + 0.36(400)(12) - (8500)(A/F, 5\%, 5)$$
$$= 6057/12$$
$$\approx \$504.75/\text{month}$$

6-20

Granny Gums has established a scholarship at the Martin College of Dentistry. She will make deposits into an endowment account that pays 12% per year based on the following schedule.

Year:	0	1	2	3	4	5	6
Deposit amount ($):	100	90	80	70	60	50	40

If the first scholarship is to be awarded 1 year after the first deposit is made and thereafter the award will be given indefinitely, what is the scholarship amount?

Solution

First find the present worth of the gradient deposits.

$P = 100 + 90(P/A, 12\%, 6) - 10(P/G, 12\%, 6) = \$380.69$

$A = 380.69(A/P, 12\%, \infty) = 380.69(.12) = \$45.68$

6-21

A proposed steel bridge has an indefinite life. The initial cost of the bridge is $3,750,000, and annual maintenance costs are estimated to be $25,000. The bridge deck will be resurfaced every 10 years for $900,000, and anticorrosion paint will be applied every 5 years for $250,000. If the interest rate is 8%, what is the EUAC?

If 650,000 axles will cross the bridge each year, what approximate toll per axle should be charged? Give your answer to the nearest nickel.

Solution

$\text{EUAC} = 3,750,000(A/P, 8\%, \infty) + 900,000(A/F, 8\%, 10) + 250,000(A/F, 8\%, 5) + 25,000$

$\qquad = \$429,725$

$\text{Toll} = 429,725/650,000 = 0.6611$

$\qquad \approx 70\cent \text{ per axle}$

6-22

A college has been willed $100,000 to establish a permanent scholarship. If funds are invested at 6% and all funds earned are disbursed yearly, what will be the value of the scholarship in the sixth year of operation?

Solution

$A = Pi = 100,000(0.06) = \$6000 \text{ for any year}$

6-23

Calculate the annual worth of a quarterly lease payment of $500 at 8% interest.

Solution

Recall that lease payments are made at beginning of period.

$i = 8/4 = 2\%$ $n = 4$

$A = 500(F/P, 2\%, 1)(F/A, 2\%, 4) = \$2102.22$

Alternate solution:

$A = [500 + 500(P/A, 2\%, 3)](F/P, 2\%, 4) = \$2101.24$

6-24

An airport expansion that is expected to be used indefinitely is under way at Jackson Hole Metro Airport. Land acquisition and major earthworks that will last as long as the airport is used are expected to cost $600 million. Terminal construction that will last 20 years is budgeted at $200 million. (Assume that the terminal is to be identically replaced every 20 years.)   Runway construction will cost $150 million. The runways will also be used indefinitely, with repaving required every 10 years at a cost of $10 million. The operating and maintenance costs are estimated to be $15 million per year. What is the annual cost of the project if $i = 5\%$ and the airport will be used indefinitely?

Solution

$\text{EUAC} = 600(A/P, 5\%, \infty) + 200(A/P, 5\%, 20) + 150(A/P, 5\%, \infty) + 10(A/F, 5\%, 10) + 15 =$ $69,335,000

6-25

Green County is planning to construct a bridge across the south branch of Carey Creek to facilitate traffic flow though Clouser Canyon. The first cost for the bridge will be $9,500,000. Annual maintenance and repairs the first year of operation, estimated to be $10,000, are expected to increase by $1000 each year thereafter. In addition to regular maintenance, every 5 years the roadway will be resurfaced at a cost of $750,000, and the structure must be painted every 3 years at a cost of $100,000. If Green County uses 5% as its cost of money and the bridge is expected to last for 20 years, what is the EUAC?

Solution

$\text{EUAC} = 9,500,000(A/P, 5\%, 20) + [10,000 + 1000(A/G, 5\%, 20)]$
$+ 750,000(A/F, 5\%, 5)(P/A, 5\%, 15)(A/P, 5\%, 20)$
$+ 100,000(A/F, 5\%, 3)(P/A, 5\%, 18)(A/P, 5\%, 20)$
$= \$922,551$

6-26

Two alternative investments are being considered. What is the minimum uniform annual benefit that will make Investment B preferable to Investment A? Assume that interest is 10%.

Year	A	B
0	−$500	−$700
1–5	+150	?

Solution

$\text{EUAW}_A = \text{EUAW}_B$

$-500(A/P, 10\%, 5) + 150 = -700(A/P, 10\%, 5) + X$

$X = \$202.76$

6-27

Consider two investments:

1. Invest $1000 and receive $110 at the end of each month for the next 10 months.
2. Invest $1200 and receive $130 at the end of each month for the next 10 months.

If this were your money, and you wanted to earn at least 12% interest on it, which investment would you make, if any? Solve the problem by annual cash flow analysis.

Solution

Alternative 1: $\text{EUAW} = \text{EUAB} - \text{EUAC} = 110 - 1000(A/P, 1\%, 10) = \$4.40$

Alternative 2: $\text{EUAW} = \text{EUAB} - \text{EUAC} = 130 - 1200(A/P, 1\%, 10) = \$3.28$

Maximum EUAW; therefore, choose Alternative 1.

6-28

Morton and Moore LLC (M^2) is trying to decide between two machines that are necessary in its manufacturing facility. If M^2 has a MARR of 10%, which of the following machines should be chosen?

	Machine A	Machine B
First cost	$45,000	$24,000
Annual operating costs	31,000	35,000
Overhaul in Years 2 and 4	—	6,000
Overhaul in Year 5	12,000	—
Salvage value	10,000	8,000
Useful life	8 years	6 years

Solution

$EUAC_A = -45,000(A/P, 10\%, 8) - 31,000 - 12,000(P/F, 10\%, 5)(A/P, 10\%, 8)$
$+ 10,000(A/F, 10\%, 8)$
$= -\$39,955$
$EUAC_B = -24,000(A/P, 10\%, 6) - 35,000 - 6000[(P/F, 10\%, 2)$
$+ (P/F, 10\%, 4)](A/P, 10\%, 6)$
$+ 8000(A/F, 10\%, 6) = -\$41,533$
Minimize EUAC; therefore, choose Machine *A*.

6-29

A land surveyor just starting in private practice needs a van to carry crew and equipment. He can lease a used van for $8000 per year, paid at the beginning of each year, in which case maintenance is provided. Alternatively, he can buy a used van for $16,000 and pay for maintenance himself. He expects to keep the van for 3 years, at which time he would sell it for an anticipated $3500. Given a MARR of 6%, what is the most the surveyor should pay for uniform annual maintenance to make it worthwhile to buy the van instead of leasing it?

Solution

Lease:

$EUAC = 6500(F/P, 6\%, 1) = 6500(1.06) = \$6890$

Buy:

$EUAC = 16,000(A/P, 6\%, 3) + M - 3500(A/F, 6\%, 3)$
Setting equal and solving for *M* yields
$6890 = 5985.60 + M - 1099.35$
$M = \$2003.75$

6-30

Assuming monthly payments, which would be the better financing plan on the same $19,000 car?
Plan A: 6% interest on the full amount for 48 months
Plan B: a $2500 rebate (discount) and 12% interest on the remaining amount for 48
   months

Solution

Plan A. $A = 19,000(A/P, \frac{1}{2}\%, 48) = \$446.50/mo.$
Plan B. $A = 16,500(A/P, 1\%, 48) = \$433.95/mo.$ Choose Plan B.

6-31

The town of Dry Hole needs an additional supply of water from Duck Creek. The town engineer has selected two plans for comparison. The gravity plan would divert water at a point 10 miles up Duck Creek and carry it through a pipeline by gravity to the town. A system using a pumping station would divert water at a point closer to town and pump it into the town. The pumping plant would be built in two stages, with 75% of its capacity installed initially and the remaining 25% installed 10 years later. The engineer has assumed that each plan will last 40 years and be worthless at the end of its life. Use the following data and an interest rate of 8% to find the maximum that should be paid for the gravity plan.

	Gravity	Pumping
Initial investment	$??????	$1,800,000
Completion cost in 10th year		350,000
Annual operating and maintenance costs	$10,000	$25,000
Annual power costs:		
Average costs the first 10 years	0	$ 50,000
Average costs the next 30 years	0	$100,000

Solution

Gravity:

$$EUAC = X(A/P, 8\%, 40) - 10,000$$
$$= 0.0839X - 10,000$$

Pumping:

$$EUAC = 1,800,000(A/P, 8\%, 40) + 350,000(P/F, 8\%, 10)(A/P, 8\%, 40) + 25,000$$
$$+ [50,000(P/A, 8\%, 10) + 100,000(P/A, 8\%, 30)(P/F, 8\%, 10)](A/P, 8\%, 40)$$
$$= \$261,522$$

Setting the two alternatives equal yields

$$0.0839X - 10,000 = 261,522$$
$$X = \$3,236,257$$

6-32

Fitzgerald, Ivy, Garcia, Nichols, Eudy, Williams, Thomas, Owens, and Nagy (FIGNEWTON) Inc. must replace its fig-crushing equipment. The alternatives under consideration are presented below.

Alternative	First Cost	Net Annual Costs	Useful Life
A	$170,500	$14,675	5 years
B	205,000	17,000	7 years
C	242,500	16,350	8 years
D	290,000	14,825	10 years

FIGNEWTON's cost of capital is 8%. Which alternative should be chosen?

Solution

$$EUAC_A = 170,500(A/P, 8\%, 5) + 14,675 = \$57,385.25$$

$$EUAC_B = 205,000(A/P, 8\%, 7) + 17,000 = \$56,380.50$$

$$EUAC_C = 242,500(A/P, 8\%, 8) + 16,350 = \$58,545.00$$

$$EUAC_D = 290,000(A/P, 8\%, 10) + 14,825 = \$58,035.00$$

Minimize EUAC; choose alternative B.

6-33

The Tennessee Department of Highways is trying to decide whether it should "hot-patch" a short stretch of an existing highway or resurface it. If the hot-patch method is chosen, approximately 500 cubic meters of material would be required at a cost of $800/cubic meter (in place). If hot-patched, the shoulders will have to be improved at the same time at a cost of $24,000. The shoulders must be maintained at a cost of $3000 every 2 years. The annual cost of routine maintenance on the patched road is estimated to be $6000.

Alternatively, the state can resurface the road at a cost of $500,000. If maintained properly, at a cost of $2000 per year beginning in the second year, the surface will last for 10 years. The shoulders would require reworking at the end of the fifth year at a cost of $15,000. Regardless of the method selected, the road will be completely rebuilt in 10 years. At an interest rate of 9%, which alternative should be chosen?

Solution

Hot-Patch:

$$EUAC = 500(800)(A/P, 9\%, 10) + 24,000(A/P, 9\%, 10)$$
$$+ 3,000(A/F, 9\%, 2)(P/A, 9\%, 8)(A/P, 9\%, 10) + 6000 = \$73,297$$

Resurface:

$$EUAC = 500,000(A/P, 9\%, 10) + 15,000(P/F, 9\%, 5)(A/P, 9\%, 10)$$
$$+ 2000(P/A, 9\%, 9)(P/F, 9\%, 1)(A/P, 9\%, 10) = \$81,133$$

Minimize EUAC; choose the hot-patch alternative.

6-34

A semiconductor manufacturer has been ordered by the city to stop discharging acidic waste liquids into the city sewer system. Your analysis shows that the company should select one of the following three systems.

System	Installed Cost	Annual Operating Cost	Salvage Value
CleanH$_2$O	$30,000	$6000	$ 2,000
Acid Free	35,000	5000	5,000
Evergreen	80,000	1000	40,000

If the system is expected to be used for 20 years and to last that long, as well, and money is worth 8%, which system should be purchased?

Solution

CleanH$_2$O: EUAC $= 6000 + 30,000(A/P, 8\%, 20) - 2000(A/F, 8\%, 20) = \$9013$

Acid Free: EUAC $= 5000 + 35,000(A/P, 8\%, 20) - 5000(A/F, 8\%, 20) = \$8456$

Evergreen: EUAC $= 1000 + 80,000(A/P, 8\%, 20) - 40,000(A/F, 8\%, 20) = \$8276$

Purchase the system with the lowest EUAC, Evergreen.

6-35

Consider Projects A and B. Which project would you approve if a project must be selected? The expected period of service is 15 years, and the interest rate is 10%.

	Project A	Project B
Initial cost	$50,000	$75,000
Annual operating costs	15,000	10,000
Annual repair costs	5,000	3,000
Salvage value	5,000	10,000

Solution

Project A:

$\text{EUAC}_A = 50,000(A/P, 10\%, 15) + 20,000 - 5000(A/F, 10\%, 15) = \$26,417.50$

Project B:

$\text{EUAC}_B = 75,000(A/P, 10\%, 15) + 13,000 - 10,000(A/F, 10\%, 15) = \$22,547.50$

Choose the least cost: Project B.

6-36

The construction costs and annual maintenance costs of two alternatives for a canal are given. a. Use equivalent uniform annual cost (EUAC) analysis to decide which alternative you would recommend. Assume 7% interest and infinite life. b. What is the capitalized cost of maintenance for the alternative you choose?

	Alternative A	Alternative B
Construction cost	$25,000,000	$50,000,000
Annual maintenance costs	3,500,000	2,000,000

Solution

a.   Alternative A: EUAC $= A + Pi = 3.5M + 25M(0.07) = \$5,250,000$
     Alternative B: EUAC $= A + Pi = 2.0M + 50M(0.07) = \$5,500,000$
     Fixed output $\therefore$ minimize cost; choose A.

b. $P = A/i = 3,500,000/0.07 = \$50,000,000$ for the maintenance costs

6-37

The manager of Cats-N-The-Pond Inc. is trying to decide between two alternative designs for an aquacultural facility. Both facilities produce the same number of fish for sale. The first alternative costs $250,000 to build and has a first-year operating cost of $110,000. Operating costs are estimated to increase by $10,000 per year for each year after the first.

The second alternative costs $450,000 to build and has a first-year operating cost of $40,000 per year, escalating at $5000 per year for each year after the first. The estimated life of both plants is 10 years, and each has a salvage value that is 10% of construction cost.

Assume an 8% interest rate. Use equivalent uniform annual cost (EUAC) analysis to determine which alternative should be selected.

Solution

	Alternative 1	Alternative 2
First cost	$250,000	$450,000
Uniform annual costs	110,000	40,000
Gradient	10,000	5,000
Salvage value	25,000	45,000

    Alternative 1: EUAC $= 250,000(A/P, 8\%, 10) - 25,000(A/F, 8\%, 10) + 110,000 +$
$10,000(A/G, 8\%, 10)$

$$= \$184,235$$

    Alternative 2: EUAC $= 450,000(A/P, 8\%, 10) - 45,000(A/F, 8\%, 10) + 40,000 + 5000(A/G,$
$8\%, 10)$

$$= \$123,300$$

Fixed output (same amount of fish for sale) $\therefore$ minimize EUAC. Choose Alternative 2.

6-38

The plant engineer of a major food processing corporation is evaluating alternatives to supply electricity to the plant. He will pay $3 million for electricity purchased from the local utility at the end of the first year and estimates that this cost will increase thereafter at $300,000 per year. He desires to know if he should build a 4000-kilowatt power plant. His operating costs (other than fuel) for such a power plant are estimated to be $130,000 per year. He is considering two alternative fuels:

a. Wood: Installed cost of the power plant is $1200/kW. Fuel consumption is 30,000 tons per year. Fuel cost for the first year is $20/ton and is estimated to increase at a rate of $2/ton for each year after the first. No salvage value.

b. Oil: Installed cost is $1000/kW. Fuel consumption is 46,000 barrels per year. Fuel cost is $34 per barrel for the first year and is estimated to increase at $1/barrel per year for each year after the first. No salvage value.

If interest is 12%, and the analysis period is 10 years, which alternative should the engineer choose? Solve the problem by equivalent uniform annual cost analysis (EUAC).

Solution

	Do Nothing	Wood	Oil
First cost	0	4000 × 1200 = 4,800,000	4000 × 1000 = 4,000,000
Annual oper. costs	0	130,000	130,000
Annual energy costs	3,000,000	30,000 × 20 = 600,000	46,000 × 34 = 1,564,000
Gradient	300,000	30,000 × 2 = 60,000	46,000 × 1 = 46,000

Do Nothing: EUAC = 3000K + 300K(A/G, 12%, 10) = $4,075,500

Wood: EUAC = 4800K(A/P, 12%, 10) + 130K + 600K + 60K(A/G, 12%, 10) = $1,794,700

Oil: EUAC = 4000K(A/P, 12%, 10) + 130K + 1564K + 46K(A/G, 12%, 10) = $2,566,190

Minimize EUAC; choose wood.

6-39

Two alternatives are being considered by a food processor for the warehousing and distribution of its canned products in a sales region. These canned products come in standard cartons of 24 cans per carton. The two alternatives are as follows.

Alternative A: To have its own distribution system. The administrative costs are estimated at $43,000 per year, and other general operating expenses are calculated at $0.009 per carton. A warehouse will have to be purchased, at a cost of $300,000.

Alternative B: To sign an agreement with an independent distribution company that is asking a payment of $0.10 per carton distributed.

Assume a study period of 10 years and that the warehouse can be sold at the end of this period for $200,000.

    a.   Which alternative should be chosen, if management expects that the number of cartons to be distributed will be 600,000 per year?

    b.   Find the minimum number of cartons per year that will make the alternative of having a distribution system (Alt. *A*) more profitable than to sign an agreement with the distribution company (Alt. *B*).

Solution

    a.   For 600,000 cartons/year

Alternative *A*:

Capital expenses = $300,000(A/P, 10\%, 10) - 200,000(A/F, 10\%, 10) = \$36,270$

	Administration		$43,000
Annual Costs:	Operating expenses	$0.009 \times 600{,}000$	5,400
		=	
	Capital expenses		36,270
		Total =	$84,670

Total annual costs = $84,670

Alternative *B*:

Total annual costs = $0.10 \times 600{,}000 = \$ 60{,}000$

∴ Sign an agreement for Alternative *B*.

    b.   Let *M* = number of cartons/year.

The EUAC for Alternative *B* (agreement) = $\text{EUAC}_{\text{AGREEMENT}} = 0.10M$

The EUAC for Alternative *A* (own system) = $\text{EUAC}_{\text{OWN}} = 43{,}000 + 0.009M + 36{,}270$

We want $\text{EUAC}_{\text{OWN}} < \text{EUAC}_{\text{AGREEMENT}}$

$$43{,}000 + 0.009M + 36{,}270 < 0.10M$$
$$79{,}270 < (0.10 - 0.009)M$$
$$79{,}270/0.091 < M$$
$$871{,}099 < M$$

∴ Owning the distribution system is more profitable for 871,100 or more cartons/year.

6-40

Dorf Motors Manufacturing must replace one of its tow motors. The net present cost of Alternative *A* is $8956, Alternative *B* is $5531, and Alternative *C* is $4078. Alternative *A* is expected to last for 12 years; Alternative *B* has an expected life of 7 years; and Alternative *C* is expected to last for 5 years. If Dorf's MARR is 5%, which Alternative (if any) should be chosen using EUAC.

Solution

Annual worth analysis is appropriate because of the different useful lives.

$EUAC_A = 8956(A/P, 5\%, 12) = \$1010.23$

$EUAC_B = 5531(A/P, 5\%, 7) = \$955.76$

$EUAC_C = 4078(A/P, 5\%, 5) = \$942.02$

Minimize EUAC; choose Alternative C.

6-41

According to the manufacturers' literature, the costs of running automatic grape peelers, if maintained according to the instruction manuals, are as follows.

	Slippery	Grater
First cost	$500	$300
Maintenance	$100 at end	Year 1 $ 0
	of Years 2,	2 50
	4, 6, and 8	3 75
		4 100
		5 125
Useful life	10 years	5 years

Which alternative is preferred if MARR = 8%?

Solution

Slippery:
$EUAC = [500 + 100(A/F, 8\%, 2)(P/A, 8\%, 8)](A/P, 8\%, 10) = \$115.67$
Grater:
$EUAC = [300 + [50(P/A, 8\%, 4) + 25(P/G, 8\%, 4)](P/F, 8\%, 1)](A/P, 8\%, 5) = \$140.52$
Therefore, choose Slippery with lower EUAC.

6-42

Two options are available for the reroofing of Pinkley Pickles Inc. The first option, a traditional tar and gravel roof, costs $75,000 and has a life of 15 years. The second option, a neoprene membrane roof, costs $85,000 and has a life of 20 years. The company expects to occupy the building for 50 years. Using an interest rate of 5% and annual worth analysis, which roof should be chosen?

Solution

T & G

EUAC = $75,000(A/P, 5\%, 15)(P/A, 5\%, 60)(A/P, 5\%, 50)$ = \$7495

Neoprene

EUAC = $85,000(A/P, 5\%, 20)(P/A, 5\%, 60)(A/P, 5\%, 50)$ = \$7075

Since the criterion is a minimized EUAC, select the neoprene roof.

Alternate solution

T & G

EUAC = $75,000[1 + (P/F, 5\%, 15) + (P/F, 5\%, 30) + (P/F, 5\%, 45)](A/P, 5\%, 50)$ = \$7495

Neoprene

EUAC = $85,000[1 + (P/F, 5\%, 20) + (P/F, 5\%, 40)](A/P, 5\%, 50)$ = \$7075

6-43

The following alternatives describe possible projects for the use of a vacant lot. In each case the project cost includes the purchase price of the land.

	Parking Lot	Gas Station
Investment cost	$50,000	$100,000
Annual income	35,000	85,000
Annual operating expenses	25,000	$70,000 in Year 1, then increasing by 1000/yr
Salvage value	10,000	10,000
Useful life	5 years	10 years

a. If the minimum attractive rate of return (MARR) equals 18%, what should be done with the land?

b. Is it possible that the decision would be different if the MARR were higher than 18%? Why or why not? (No calculations necessary.)

Solution

a. $\text{EUAW}_{P.L.} = (35,000 - 25,000) - 50,000(A/P, 18\%, 5) + 10,000(A/F, 18\%, 5) = -\$4592$

$\text{EUAW}_{G.S.} = (85,000 - 70,000) - 100,000(A/P, 18\%, 10) + 10,000(A/F, 18\%, 10) -$
$\qquad 1000(A/G, 18\%, 10)$
$\qquad = -\$10,019$

Since both EUAWs are negative, leave the lot vacant.

b. No. Higher MARR favors lower first-cost projects, and the lowest first-cost project (do nothing) has already been chosen.

6-44

Given the following information about possible investments being considered by the ABC Block Company, what is the best choice at a minimum attractive rate of return (MARR) of 10%?

	A	B
Investment cost	$5000	$8000
Annual benefits	1200	800
Useful life	5 years	15 years

Solution

Since the useful lives are different, use equivalent annual worth analysis.

$EUAW_A = 1200 - 5000(A/P, 10\%, 5) = -\$119.00$

$EUAW_B = 800 - 8000(A/P, 10\%, 15) = -\$252.00$

Choose the do-nothing alternative.

6-45

You are considering purchasing the Press-o-Matic or Steam-It-Out automatic ironing system to allow you to handle more dry cleaning business. Both machines have the same cost, $5000. The Press-o-Matic will generate a positive cash flow of $1300 per year for 5 years and then be of no service or salvage value. The Steam-It-Out will generate a positive cash flow of $800 per year for 10 years and then be of no service or salvage value. You plan to be in the dry cleaning business for the next 10 years. How would you invest the $5000 you have in your hand if you feel the time value of money is worth the same as your high-interest bank account offers, which is

a. 8%?
b. 12%?

Solution

a, Press EUAW $= 1300 - 5000(A/P, 8\%, 5) = \$47.50$
    Steam EUAW $= 800 - 5000(A/P, 8\%, 10) = \$55.00$
    Choose the higher EUAW, Steam-It-Out.

b. Press EUAW $= 1300 - 5000(A/P, 12\%, 5) = -\$87.00$
    Steam EUAW $= 800 - 5000(A/P, 12\%, 10) = -\$85.00$
    Choose neither option because both have a negative annual worth.

6-46

Data for Machines X and Y are listed. With an interest rate of 8%, and based upon equivalent uniform annual cost (EUAC), which machine should be selected?

	X	Y
First cost	$5000	$10,000
Annual maintenance	500	200
Salvage value	600	1,000
Useful life	5 years	15 years

Solution

Machine X:

EUAC = 5000(A/P, 8%, 5) – 600(A/F, 8%, 5) + 500 = $1650.20

Machine Y:

EUAC = 10,000(A/P, 8%, 15) – 1000(A/F, 8%, 15) + 200 = $1331.20

Decision criterion is minimize EUAC; therefore, choose Y.

6-47

Assuming a 10% interest rate, determine which alternative should be selected.

	A	B
First cost	$5300	$10,700
Uniform annual benefit	1800	2,100
Salvage value	0	200
Useful life	4 years	8 years

Solution

Alternative A:

EUAW = 5300(A/P, 10%, 4) – 1800 = $127.85

Alternative B:

EUAW = 10,700(A/P, 10%, 8) + 200(A/F, 10%, 8) – 2100 = $112.30

Choose alternative A.

6-48

A company must decide whether to buy Machine A or Machine B. After 5 years, Machine A will be replaced with another A.

	Machine A	Machine B
First cost	$10,000	$20,000
Annual maintenance	1,000	0
Salvage value	10,000	10,000
Useful life	5 years	10 years

With the minimum attractive rate of return (MARR) = 10%, which machine should be purchased?

Solution

$EUAW_A = -10,000(A/P, 10\%, 5) - 1000 + 10,000(A/F, 10\%, 5) = -\$2000$

$EUAW_B = -20,000(A/P, 10\%, 10) + 10,000(A/F, 10\%, 10) = -\$2627$

Therefore, Machine A should be purchased.

6-49

Your company bought equipment for $25,000 by paying 10% down and the rest on credit. The credit arrangement is for 5 years at 7% interest. After making 20 payments, you want to pay off the loan. How much do you owe?

Solution

The present value of the loan is based on the remaining payments.

	A	B	C	D	E	F	G	H	I	J
1	Problem	i	N	PMT	PV	FV	Solve for	Answer	Formula	
2	6-49	0.583%	60		-$22,500	0	PMT	$445.53	Payment	=PMT(B2,C2,E2,F2)
3										
4		0.583%	40	-$445.53		0	PV	$15,853.64	Amt. Owed	=PV(B4,C4,D4,F4)

You owe $15,853.64.

6-50

A personal loan for $20,000 has an interest rate of 12% over 5 years with monthly payments. What is the monthly uniform payment? What is owed on the loan after 3 years?

Solution

The present value of the loan is based on the remaining payments.

	A	B	C	D	E	F	G	H	I
1	Problem	i	N	PMT	PV	FV	Solve for	Answer	Formula
2	6-50	1.00%	60		$20,000	0	PMT	-444.89	=PMT(B2,C2,E2,F2)
3									
4		1.00%	24	-$444.89		0	PV	9,450.95	=PV(B4,C4,D4,F4)

You owe $9,450.95.

6-51

You are interested in leasing a car for $425 per month, due at the beginning of each month. Using an interest rate of 4% annually, what is the present worth of a one-year lease for this car?

Solution

	A	B	C	D	E	F	G	H	I
1	Problem	i	N	PMT	PV	FV	Solve for	Answer	Formula
2	6-51	0.333%	12	-425.00		0	PV	$5,007.83	=PV(B2,C2,D2,F2,1)

The present worth is $5007.83

6-52

Your uncle paid $300,000 for an annuity that will pay $2200 per month for life. The insurance company that sold the annuity says that they are paying interest of 4%. How long (in months) do they plan on paying the annuity?

Solution

	A	B	C	D	E	F	G	H	I
1	Problem	i	N	PMT	PV	FV	Solve for	Answer	Formula
2	6-52	0.333%		-2,200	300,000	0	NPER	182.14	=NPER(B2,D2,E2,F2)

They are planning on paying for 182.14 months, or just over 15 years.

Chapter 7

Rate of Return
Analysis

7-1

Andrew T. invested $15,000 in a high-yield account. At the end of 30 years he closed the account and received $539,250. Compute the effective interest rate he received on the account.

Solution

Recall that $F = P(1 + i)^n$

$$539{,}250 = 15{,}000(1 + i)^{30} \implies 539{,}250/15{,}000 = (1 + i)^{30}$$

$$35.95 = (1 + i)^{30}$$

$$\sqrt[30]{35.95} = 1 + i$$

$$1.1268 = 1 + i$$

$$0.1268 = i$$

$$i = 12.68\%$$

7-2

The heat loss through the exterior walls of a processing plant is expected to cost the owner $3000 next year. A salesman from Superfiber, Inc. claims he can reduce the heat loss by 80% with the installation of $15,000 worth of Superfiber now. If the cost of heat loss rises by $200 per year, after next year (gradient), and the owner plans to keep the building 10 years, what is his rate of return, neglecting depreciation and taxes?

Solution

NPW = 0 at the rate of return

$$0 = -15{,}000 + 0.8(3000)(P/A, i\%, 10) + 0.8(200)(P/G, i\%, 10)$$

Try 12%: $1800.64
Try 15%:   −$237.76
By interpolation, $i = 14.7\%$.

Spreadsheet Solution

	A	B
1	10	years
2	80%	factor
3	$ 15,000	installation cost
4	$ 3,000	cost first year
5	$ 200	Gradient
6		
7	Year	Cash Flow
8	0	-15,000
9	1	2,400
10	2	2,560
11	3	2,720
12	4	2,880
13	5	3,040
14	6	3,200
15	7	3,360
16	8	3,520
17	9	3,680
18	10	3,840
19		
20	IRR	14.62%

7-3

Does the following project have a positive or negative rate of return? Show how this is known to be true.

Investment cost	$2500
Net benefits	300 in Year 1, increasing by $200 per year
Salvage value	50
Useful life	4 years

Solution

Year	Benefits	
1	300	Total benefits are less than investment,
2	500	so the "return" is negative.
3	700	
4	900	
4	50	
	Total = $2450 < Cost	

7-4

At what interest rate would $1000 at the end of 2020 be equivalent to $2000 at the end of 2027?

Solution

$(1 + i)^7 = 2$

$i = (2)^{1/7} - 1 = 0.1041 = 10.41\%$

7-5

A piece of art, purchased three months ago for $12,000, has just been sold for $15,000. What nominal annual rate of return did the seller receive on her investment?

Solution

$i = (15,000 - 12,000)/12,000 = 0.25 = 25\%$ over 3 months (one quarter of a year)

$r = 25\% \times 4 = 200\%$ nominal annual

7-6

Some time ago a young engineer obtained a mortgage at a 12% interest rate, for a total of $102,000. She has to pay 240 more monthly payments of $1049.19. As interest rates are going down, she inquires about the conditions under which she could refinance the mortgage. If the bank charges an origination fee of 2% of the amount to be financed, and if the bank and the engineer agree that the fee will be paid by combining the fee with the refinanced mortgage, what percentage rate would make refinancing her mortgage attractive, if the new mortgage terms require 120 payments?

Solution

The amount to be refinanced:

$i = 12/12 = 1\%$

PW of 240 monthly payments left = 1049.19(*P/A*, 1%, 240)

$= \$95,286.24$

New loan fee (2%) = 95,286.24(0.02) = $1905.72

$\Rightarrow$ Total amount to refinance = 95,286.24 + 1905.72 = $97,191.96

The new monthly payments are: $A_{NEW} = 97,191.96(A/P, i, 120)$

The current payments are: $A_{OLD} = 1049.19$

We want $A_{NEW} < A_{OLD}$.

Substituting $\Rightarrow$ 97,191.96 (*A/P*, *i*, 120) < 1049.19

$(A/P, i, 120) < 1049.19/97,191.96 = 0.0108$

$(A/P, \frac{1}{4}\%, 120) = 0.00966$

$(A/P, \frac{1}{2}\%, 120) = 0.0111$

$\frac{1}{4}\% < i < \frac{1}{2}\%$ $\therefore$ interpolate

$i = 0.4479\%$

This corresponds to a nominal annual percentage rate of $12 \times 0.4479 = 5.375\%$.

Therefore, the engineer must wait until interest rates are less than 5.375%.

Spreadsheet Solution

	A	B	C	D	E	F	G	H	I
1	Problem	i	n	PMT	PV	FV	Solve for	Answer	Formula
2	7-6								
3	PW of 240 monthly payments left								
4		1.00%	240	1049.19		0	PV	-$95,286.82	=PV(B4,C4,D4,F4)
5	Fee				-1905.74				
6	Refinance				-97,192.56				
7									
8	Highest attractive interest makes new payment = old payment								
9			120	1049.19	-97,192.56	0	RATE	0.449%	=RATE(C8,D8,E8,F8)
10								nominal	5.38%	=+H8*12

7-7

A 9.25% coupon bond issued by Gurley Gears LLC is purchased January 1, 2020, and matures December 31, 2028. The purchase price is $1079 and interest is paid semiannually. If the face value of the bond is $1000, determine the effective internal rate of return.

Solution

$n = 2 \times 9 = 18$ ½-year periods

each interest payment of the bond = $(0.5)(0.0925)(1000) = \$46.25$

½ Year		4%
0	First cost	−1079.00
1–18	Interest 46.25(P/A, i, 18)	585.48
18	Maturity 1000(P/F, i, 18)	493.60
	NPW =	$0.08

$IRR = (1 + 0.04)^2 - 1$
$= 8.16\%$

Spreadsheet Solution

	A	B	C	D	E	F	G	H	I
1	Problem	i	n	PMT	PV	FV	Solve for	Answer	Formula
2	7-7		18	46.25	-1079	1000	RATE	4.00%	=RATE(C2,D2,E2,F2)
3								nominal	8.00%	=H2*2
4								effective	8.16%	=EFFECT(H3,2)

7-8

Sain and Lewis Investment Management (SLIM), Inc. is considering the purchase of a number of bonds to be issued by Southeast Airlines. The bonds have a face value of $10,000 with an interest rate of 7.5% payable annually. The bonds will mature 10 years after they are issued. The issue price is expected to be $8750. Determine the yield to maturity (IRR) for the bonds. If SLIM Inc. requires at least a 10% return on all investments, should the firm invest in the bonds?

Solution

Year		10%	9%
0	First cost	−8750.00	−8750.00
1–10	Interest = 750(P/A, i, 10)	4608.75	4813.50
10	Maturity = 10,000(P/F, i, 10)	3855.00	4224.00
	NPW =	−$286.25	$287.50

9% < IRR < 10%   Interpolating, $i = 9.5\% \Rightarrow$ Do not invest.

7-9

A bond with a face value of $1000 can be purchased for $800. The bond will mature 5 years from now, and the bond dividend rate is 6%. Dividends are paid every 6 months. What effective interest rate would an investor receive if she purchased the bond?

Solution

NPW = 0 at the rate of return

$0 = -800 + 30(P/A, i, 10) + 1000(P/F, i, 10)$

Try $i = 5\%$: NPW = $45.56

Try $i = 6\%$: NPW = −$20.80

5% < i < 6%   Interpolating, $i = 5.69\%$ per half year

Effective interest rate = $(1 + 0.0569)^2 - 1 = 0.1170 = 11.70\%$

Spreadsheet Solution

	A	B	C	D	E	F	G	H	I
1	Problem	i	n	PMT	PV	FV	Solve for	Answer	Formula
2	7-9		10	30	-800	1000	RATE	5.68%	=RATE(C2,D2,E2,F2)
3							Nominal	11.35%	=+H2*2
4							Effective	11.67%	=EFFECT(H3,2)

7-10

You find a car you like that costs $18,000. The dealer is offering 0% financing for 4 years, but you must pay a financing fee of $800.

a. What is the effective annual interest of this "free financing" deal?

b. If you get a cash discount and can pay $17,250, what is the effective annual interest rate of the financing?

   a. Monthly payment = (18,000 + 900) / 48 = 393.75

   18,000 = 393.75(P/A, i, 48)

   (P/A, i, 48) = 18,000 / 393.75 = 45.714

   (P/A, 0.25%, 48) = 45.179

   (P/A, 0%, 48) = 48

   Interpolating, $i = 0.20\%$ monthly

   Effective annual = $(1 + 0.0020)^{12} - 1 = 0.0243 = 2.43\%$

b. $17,250 = 393.75(P/A, i, 48)$, or $(P/A, i, 48) = 17,250 / 393.75 = 43.810$

$(P/A, 0.25\%, 48) = 45.179$

$(P/A, 0.5\%, 48) = 42.580$

Interpolating, $i = 0.382\%$ monthly

Effective annual $= (1 + 0.00382)^{12} - 1 = 0.0468 = 4.68\%$

Spreadsheet Solution

	A	B	C	D	E	F	G	H
1	Problem	i	n	PMT	PV	FV	Solve for	Answer
2	7-10a	0	48		18,900	0	PMT	$393.75
3			48	-$393.75	18,000	0	RATE	0.20%
4							Effective	2.44%
5								
6	7-10b		48	-$393.75	17,250	0	RATE	0.38%
7							Effective	4.65%

7-11

Processing equipment costs $170,000. The manufacturer offers financing at 5% interest for 5 years. If your company pays cash, the manufacturer will offer a 6% decrease in price. What effective interest rate would you be paying for the financing?

Solution

Cash cost: $(0.95)(170,000) = \$161,500$

Monthly payment $= 170,000(A/P, 6\%/12, 60) = 170,000(0.0193) = \$3281$

$161,500 = 3281(P/A, i, 60)$

$(P/A, i, 60) = 161,500 / 3281 = 49.223$

$(P/A, 0.5\%, 60) = 51.726$

$(P/A, 0.75\%, 60) = 48.174$

Interpolating, $i = 0.676\%$ per month

Effective annual $= (1 + 0.00676)^{12} - 1 = 0.0842 = 8.42\%$

Spreadsheet Solution

	A	B	C	D	E	F	G	H
1	Problem	i	n	PMT	PV	FV	Solve for	Answer
2	7-11		60	-$3,287	$161,500		RATE	0.680%
3							Effective	8.47%

7-12

Find the rate of return for a $10,000 investment that will pay $1000 per year for 20 years.

Solution

$$10,000 = 1000(P/A, i, 20)$$
$$(P/A, i, 20) = 10$$
$$(P/A, 7\%, 20) = 10.549$$
$$(P/A, 8\%, 20) = 9.818$$
$$7\% < i < 8\% \quad \text{Interpolating, } i = 7.77\%$$

7-13

Your company has been presented with an opportunity to invest in a project that is summarized as follows.

Investment required	$60,000,000
Annual operating income	14,000,000
Annual operating costs	5,500,000
Salvage value after 10 years	0

The project is expected to operate as shown for 10 years. If your management requires a return of 6% on its investments before taxes, would you recommend this project based on rate of return analysis?

Solution

$$\text{Net income} = 14,000,000 - 5,500,000$$
$$= \$8,500,000$$
$$\text{NPW} = 0 \text{ at the rate of return}$$
$$0 = -60,000,000 + 8,500,000(P/A, i, 10)$$
$$(P/A, i, 10) = 60/8.5$$
$$= 7.0588$$
$$(P/A, 6\%, 10) = 7.360$$
$$(P/A, 7\%, 10) = 7.024$$
$$\text{Interpolating, } i = 6.9\% \qquad \text{IRR} > 6\,\%, \text{ so recommend the project.}$$

7-14

An investment that cost $15,000 is sold after 5 years for $18,917. What is the nominal rate of return on the investment, assuming annual compounding?

Solution

$F = P(F/P, i, 5)$

$18,917 = 15,000(F/P, i, 5)$

$(F/P, i, 5) = 18,917/15,000 = 1.2611$

$(F/P, 4\frac{1}{2}\%, 5) = 1.246$

$(F/P, 5\%, 5) = 1.276$

$4\frac{1}{2}\% < i < 5\%$ Interpolating, $i = 4.75\%$

7-15

Isabella made an initial investment of $5000 in a trading account with a stock brokerage house. After a period of 17 months, the value of the account had increased to $6400. Assuming that there were no additions or withdrawals from the account, what was the nominal annual interest rate earned on the initial investment?

Solution

$F = P(F/P, i, 17)$

$F/P = 6400/5000 = 1.28$

$(1 + i)^{17} = 1.28$

$1 + i = (1.28)^{1/17}$

$i = 0.0146 = 1.46\%$ monthly

Nominal annual interest rate $= 1.46\% \times 12 = 17.52\%$.

7-16

Whiplash Airbags has been presented the investment opportunity summarized as follows.

Year	0	1	2	3	4	5	6	7	8
Cash flow (1000s)	$(440)	20	40	60	80	100	120	140	160

Determine the IRR for the proposed investment.

Solution

NPW = 0 at the rate of return

$0 = 440,000 + 20,000(P/A, i, 8) + 20,000(P/G, i, 8)$

Try $i = 8\%$: $= \$31,060$

Try $i = 10\%$: $= -\$12,720$

Interpolating, $i = 9.42\%$

7-17

You have a choice of $2000 now or $250 now with $80 a month for 2 years. What interest rate will make these choices comparable?

Solution

$2000 = 250 + 80(P/A, i, 24)$

$P/A = 21.875$

$(P/A \; \frac{3}{4}\%, 24) = 21.889$

$(P/A \; 1\%, 24) = 21.243$

Interpolating, $i = 0.7554\%$ per month, or 9.07% per year

7-18

Tri-State Tire is considering the purchase of new inflation equipment for its Martin operation. From the following cash flows associated with the new equipment, determine the IRR.

Year	Cash Flow
0	$(2000)
1	1000
2	750
3	500
4	250
5	0
6	−250

Solution

NPW = 0 at the rate of return

$0 = -2,000 + 1000(P/A, i, 6) - 250(P/G, i, 6)$

Try $i = 7\%$: $= \$20.51$

Try $i = 8\%$: $= -\$7.35$

Interpolating, $i = 7.73\%$

7-19

One share of Harris, Andrews, and Tatum (HAT) Enterprises was purchased 5 years ago for $10.89. Dividends of 5¢ were paid each quarter over the 5-year period of ownership. The share is sold today for $18.31. Determine the effective rate of return on the stock.

Solution

$n = 4 \times 5 = 20$ ¼-year periods

¼ Year		2%	4%
0	First cost	−10.89	−10.89
1–20	Dividend .05(P/A, i, 20)	0.82	0.68
20	Sale 18.31(P/F, i, 20)	12.32	8.36
	NPW =	$2.25	−$1.85

Interpolating, $i = 3.1\%$ per period

Effective interest rate $= (1 + 0.031)^4 - 1 = 12.99\%$

7-20

If a MARR of 12% is required, which alternative should be chosen?

Year	X	Y
0	−10,000	−10,000
1	−6,000	4,000
2	8,000	4,000
3	8,000	4,000
4	8,000	4,000

Solution

The alternatives are listed in order of increasing cost for clear incremental analysis.

	A	B	C	D
1	Year	Y	X	X-Y
2	0	-10,000	-10,000	0
3	1	4,000	-6,000	-10,000
4	2	4,000	8,000	4,000
5	3	4,000	8,000	4,000
6	4	4,000	8,000	4,000
7				
8	IRR	21.86%	16.94%	9.70%

The IRR for Y is 21.86%, and the IRR for X is 16.94%, so both qualify. The incremental (X−Y) IRR is 9.70%, and is below the MARR. The added investment for X is not justified. Choose Y.

7-21

Water purification facilities are in the planning stage, with expected lives of 10 years. Two final plans are being compared using a MARR of 15%. Which alternative is preferred using incremental IRR?

	Alt. A	Alt. B
First cost	$1,200,000	$800,000
O&M cost	30,000	20,000
Annual benefits	350,000	250,000
Salvage value	50,000	−20,000

Solution

The alternatives are listed in order of increasing cost for clear incremental analysis.

◢	A	B	C	D
1		B	A	A-B
2	First cost	-800,000	-1,200,000	-400,000
3	O&M cost	-20,000	-30,000	-10,000
4	Annual benefits	250,000	350,000	100,000
5	Salvage value	-20,000	50,000	70,000
6				
7	RATE	25.79%	23.57%	19.19%

Alternatives B and A both meet the MARR. Incremental analysis shows that A justifies the increased first cost. Choose A. Hint: a "guess" may be needed to solve for the RATE.

7-22

If the firm's MARR is 10%, which alternative should be chosen assuming identical replacement?

	A	B
First cost	$10,000	$17,500
Uniform Annual Benefit	3,500	4000
Useful life, in years	5	10

Solution

	A	B	C	D
1	Year	A	B	B-A
2	0	-$10,000	-$17,500	-$7,500
3	1	$3,500	$4,000	$500
4	2	$3,500	$4,000	$500
5	3	$3,500	$4,000	$500
6	4	$3,500	$4,000	$500
7	5	-$6,500	$4,000	$10,500
8	6	$3,500	$4,000	$500
9	7	$3,500	$4,000	$500
10	8	$3,500	$4,000	$500
11	9	$3,500	$4,000	$500
12	10	$3,500	$4,000	$500
13	ROR	22.1%	18.8%	14.9%

Alternative A and B both meet the MARR. Incremental analysis shows that B justifies the increased first cost because its IRR exceeds the MARR. Choose B.

7-23

Lena wants to buy a small microwave oven, but wants the best deal. Two models are available at the discount store. A higher investment is expected to return at least 20% per year. Which should she buy?

	Nuke	Zap
First cost	$45	$80
Expected life	1 year	2 years

Solution

	A	B	C	D
1	Year	Nuke	Zap	Difference
2	0	-45	-80	-35
3	1	-45	0	45
4	2	0	0	0
5	IRR			29%

She should buy Zap because it provides a 29% rate of return, exceeding her minimum.

Appendix 7A

Difficulties in Solving for an Interest Rate

7A-1

Do these cash flows have a unique IRR?

Year	Cash Flow
0	−$20,000
1	10,000
2	−8,000
3	12,000
4	20,000

Solution

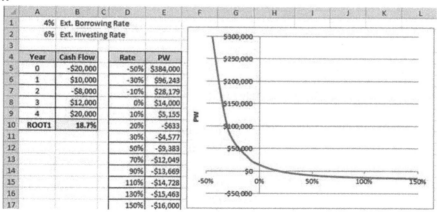

	A	B	C	D	E
1	4%	Ext. Borrowing Rate			
2	6%	Ext. Investing Rate			
3					
4	Year	Cash Flow		Rate	PW
5	0	-$20,000		-50%	$384,000
6	1	$10,000		-30%	$96,243
7	2	-$8,000		-10%	$28,179
8	3	$12,000		0%	$14,000
9	4	$20,000		10%	$5,155
10	ROOT1	18.7%		20%	-$633
11				30%	-$4,577
12				50%	-$9,383
13				70%	-$12,049
14				90%	-$13,669
15				110%	-$14,728
16				130%	-$15,463
17				150%	-$16,000

7A-2

Do these cash flows have a unique IRR? Should the project be built if a firm's MARR is 25%?

Year	Cash Flow
0	−$1250
1	2000
2	−400

Solution

There are two sign changes, so up to two roots are possible. The graphed PW shows that there are two roots, but one is negative (-80.1%) and can be ignored. Build since the IRR of 36.6% is well above the 25% MARR. The MIRR is lower since a 10% external investing rate is assumed.

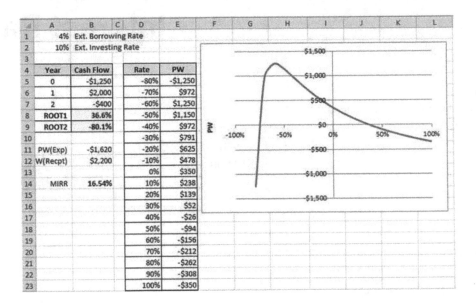

	A	B	C	D	E
1	4%	Ext. Borrowing Rate			
2	10%	Ext. Investing Rate			
3					
4	Year	Cash Flow		Rate	PW
5	0	-$1,250		-80%	-$1,250
6	1	$2,000		-70%	$972
7	2	-$400		-60%	$1,250
8	ROOT1	36.6%		-50%	$1,150
9	ROOT2	-80.1%		-40%	$972
10				-30%	$791
11	PW(Exp)	-$1,620		-20%	$625
12	W(Recpt)	$2,200		-10%	$478
13				0%	$350
14	MIRR	16.54%		10%	$238
15				20%	$139
16				30%	$52
17				40%	-$26
18				50%	-$94
19				60%	-$156
20				70%	-$212
21				80%	-$262
22				90%	-$308
23				100%	-$350

7A-3

Do these cash flows have a unique IRR? Should the project be built if a firm's MARR is 25%? If a MIRR is needed use 4% as the borrowing rate and 10% as the investing rate.

Year	Cash Flow
0	-$9000
1	8000
2	5000
3	-6000

Solution

There are two sign changes, so up to two roots are possible. The PW curve shows that there are no roots. With all PW < 0 this project should not be done. No roots exist. This is why the IRR functions return #NUM in the worksheet cells.

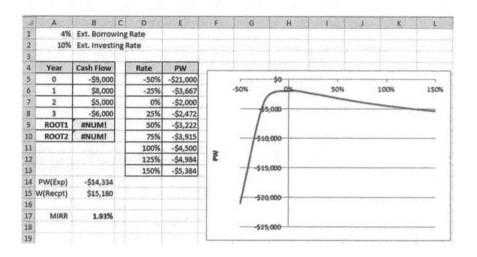

7A-4

Do these cashflows have a unique IRR? Should the project be built if a firm's MARR is 25%? If a MIRR is needed use 4% as the borrowing rate and 10% as the investing rate.

Year	Cash Flow
0	−$2000
1	7200
2	−8500
3	3300

Solution

Three roots so no unique IRR. Don't build since MIRR is only 6.81% well below MARR of 25%. Note that local minimum of −$2 and local maximum of $11 are basically 0 when compared with cash flows to the nearest $100. A PW of $9.60 at 25% does *not* imply the project should be built.

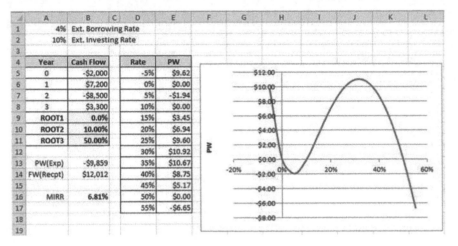

7A-5

Find modified internal rate of return (MIRR) if a firm finances money at 2% and invests money at 8%.

Year	Cash Flow
0	−$1000
1	250
2	−250
3–5	425

Solution

$$\text{PW}_{expenses} = -\$1000 - 250(P/F, 2\%, 2) = -1240$$
$$\text{FW}_{receipts} = 250(F/P, 8\%, 1) + 425(F/A, 8\%, 3) = 1650$$

Find the MIRR that will make the PW and the FW equivalent.

$$0 = \text{PW}(1 + \text{MIRR})^n + \text{FW}$$
$$0 = (-1240)(1 + \text{MIRR})^5 + 1650$$
$$(1 + \text{MIRR})^5 = 1650/1240$$
$$1 + \text{MIRR} = (1650/1240)^{1/5}$$
$$\text{MIRR} = 5.88\%$$

7A-6

Find the modified internal rate of return (MIRR) if Fly-By-Night Shipping finances money at 4% and invests money at 10%.

Year	Cash Flow
0	$-800,000
1	-150,000
2	250,000
3	-100,000
4	400,000
5–10	250,000

Solution

Find the PW of expenses using the financing rate.

$PW = -800,000 - 150,000(P/F, 4\%, 1) - 100,000(P/F, 4\%, 3) = -1,033,125$

Find the FW of receipts using the investing rate.

$FW = 250,000(F/P, 4\%, 8) + 400,000(F/P, 4\%, 6) + 250,000(F/A, 4\%, 6) = 2,506,545$

Find the MIRR that will make the PW and the FW equivalent.

$$0 = PW(1 + MIRR)^n + FW$$
$$0 = (-1,033,125)(1 + MIRR)^{10} + 2,506,545$$
$$(1 + MIRR)^{10} = 2,506,545/1,033,125$$
$$1 + MIRR = (2,506,545/1,033,125)^{1/10}$$
$$MIRR = 9.27\%$$

7A-7

Find the modified internal rate of return (MIRR) if a firm finances money at 5% and invests money at 10%.

Year	Cash Flow
0	-$10,000
1	-25,000
2–7	12,000
8	-14,250

Solution

$PW_{expenses} = -\$10,000 - 25,000(P/F, 5\%, 1) - 14,250(P/F, 5\%, 8) = -43,454$

$FW_{receipts} = 12,000(F/A, 10\%, 6)(F/P, 10\%, 1) = 101,846$

Find the MIRR that will make the PW and the FW equivalent.

$$0 = PW(1 + MIRR)^n + FW$$
$$0 = (-43,454)(1 + MIRR)^8 + 101,846$$
$$(1 + MIRR)^8 = 101,846/43,454$$
$$1 + MIRR = (101,846/43,454)^{1/8}$$
$$MIRR = 11.23\%$$

Chapter 8

Choosing the Best Alternative

8-1

Two mutually exclusive investment projects have been presented to your company. Project A's rate of return is 6%, and Project B's rate of return is 8%. The cost of Project A is less than the cost of Project B. Incremental analysis yields a rate of return of 4%. If both alternatives have the same useful life and the MARR is 5%, which project should be chosen?

Solution

Since $ROR_{B-A} <$ MARR, the increased cost of B is not justified. Choose A.

8-2

EA Construction must replace a piece of equipment. Cat and Volvo are the two best alternatives. Both alternatives are expected to last 6 years. If EA has a MARR of 11%, which alternative should be chosen? Use IRR analysis.

	Cat	Volvo
First cost	$160,000	$225,000
Annual operating cost	30,000	17,500
Salvage value	20,000	40,000

Solution

	Cat	Volvo	Increment
First cost	−$160,000	−$225,000	−65,000
Annual Op cost	−30,000	−17,500	12,500
Salvage	20,000	40,000	20,000

$IRR_{Volvo-Cat}$

NPW = 0 at the internal rate of return

$$0 = -65,000 + 12,500(P/A, i\%, 6) + 20,000(P/F, i\%, 6)$$

At $i = 10\%$: $= \$727.50$
At $i = 12\%$: $= -\$3480.50$

Interpolating: IRR = 10.35%, which is less than MARR (11%)
The increased first cost is not justified; choose the less expensive: Cat.

8-3

Horizon Wireless must rebuild a cell tower. A tower made of normal steel (NS) will cost $30,000 to construct and should last 15 years. Maintenance will cost $1000 per year. If corrosion-resistant steel (CRS) is used, the tower will cost $36,000 to build, but the annual maintenance cost will be reduced to $250 per year. Determine the IRR of building the corrosion-resistant tower. If Horizon requires a return of 9% on its capital projects, which tower should be chosen?

Solution

	Normal Steel	Corrosion Res.	Increment
Initial cost	−$30,000	−$36,000	−$6000
Maintenance	−1000	−250	+750

	A	B	C	D
1		Normal Steel	Corrosion Res.	Increment
2	Initial cost	-$30,000	-$36,000	-$6,000
3	Maintenance	-1,000	-250	750
4	Life	15	15	
5	MARR	9%		
6				
7	RATE		9.13% =RATE(+B4,D3,D2)	

The rate of return is 9.13%, exceeding the MARR and justifying the higher capital of the corrosion resistant tower.

8-4

Gamma Inc. must replace its grinding machine. Relevant information about the three "best" alternatives is given below. Each alternative has a 10-year life. Gamma's MARR is 8%. Using ROR analysis, which alternative should be selected?

	A	*B*	*C*
First cost	$30,000	$35,000	$28,000
Annual costs	2,900	2,200	3,250

Solution

Incremental analysis is required for IRR. First, organize the projects with increasing first cost.

⬦	A	B	C	D	E	F
1		10 Life				
2		8% MARR				
3						
4		C	A	B	A-C	B-A
5	First cost	-$28,000	-$30,000	-$35,000	-$2,000	-$5,000
6	Annual cost	-3,250	-2,900	-2,200	$350	$700
7				RATE	11.7%	6.6%
8					Choose A	Choose A

Determine the incremental IRR of the first set, $A-C$. The IRR of 11.7% exceeds the MARR, so the higher cost project is justified. Choose A.

Next, determine the incremental IRR of the second set, $B-A$. The IRR of 6.6% does not exceed the MARR, so the higher cost project is not justified. Choose A.

8-5

Free-Flow Sanitation Inc. is considering the purchase of advanced software. Four software packages are under consideration. Relevant information for each package is given below. Each alternative has a projected 8-year life and the MARR for the company is 6%. Using ROR analysis, which alternative should be selected?

	A	B	C	D
First cost	$5,000.00	$7,000.00	$4,000.00	$6,500.00
Annual benefits	1,842.25	2,161.40	1,640.50	1,914.70
Annual costs	1,068.62	849.31	835.35	949.31

Solution

First, organize the projects with increasing first cost.

▲	A	B	C	D	E
1		8	Life		
2		6%	MARR		
3			*C*	*A*	*D*	*B*
4	First cost		-$4,000.00	-$5,000.00	-$6,500.00	-$7,000.00
5	Annual benefits		1,640.50	1,842.25	1,914.70	2,161.40
6	Annual costs		-835.35	-1,068.62	-949.31	-849.31
7	Annual cash flow		805.15	773.63	965.39	1,312.09
8						
9			*A–C*	*D–C*	*B–C*	
10	First cost		-$1,000.00	-$2,500.00	-$3,000.00	
11	Annual net		-31.52	160.24	506.94	
12						
13		RATE	none	-12.8%	7.2%	
14			Choose C	Choose C	Choose B	

Determine the incremental IRR of the first set, $A–C$. The investment returns a negative cash flow, so there is no return. Choose the lower cost project, C.

Next, determine the incremental IRR of the next set, $D–C$. The IRR is -12.8%, which does not justify the higher investment in D. Choose C.

Last, determine the incremental IRR of the third set, $B–C$. The IRR of 7.2% exceeds the MARR, so the higher cost project is justified. Choose B.

8-6

Barber Brewing is considering investing in one of the following opportunities.

	A	*B*	*C*	*D*
First cost	$100.00	$130.00	$200.00	$330.00
Annual income	100.00	90.78	160.00	164.55
Annual cost	73.62	52.00	112.52	73.00

Each alternative has a 5-year useful life. Use net present worth to prepare a choice table.

Solution

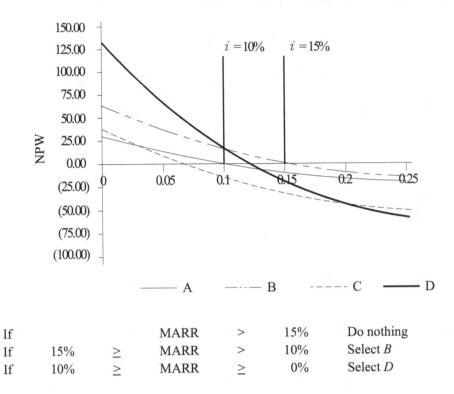

If			MARR	>	15%	Do nothing
If	15%	≥	MARR	>	10%	Select *B*
If	10%	≥	MARR	≥	0%	Select *D*

8-7

Abby W. is considering the following mutually exclusive investment projects.

	A	*B*	*C*	*D*
E				
First cost	$100.00	$130.00	$200.00	$330.00	Do
Annual income	150.00	130.78	185.00	184.55	Nothing
Annual cost	123.62	92.00	137.52	93.00	

Each alternative has a 5-year useful life. If Abby requires at least a 10% return on her investments, which alternative should she select?

Solution

First, organize the projects with increasing first cost.

▲	A	B	C	D	E	F
1		5	Life			
2		10%	MARR			
3						
4			*E*	*A*	*B*	*C*	*D*
5	First cost		0.00	-100.00	-130.00	-200.00	-330.00
6	Annual income		0.00	150.00	130.78	185.00	184.55
7	Annual cost			-123.62	-92.00	-137.52	-93.00
8							
9	Cash Flow		0	26.38	38.78	47.48	91.55
10	IRR		0%	10.0%	15.0%	6.0%	12.0%
11			Eliminate			Eliminate	
12							
13			*B−A*	*D−B*			
14	First cost		-$30.00	-$200.00			
15	Annual net		12.40	52.77			
16							
17		RATE	30.3%	10.0%			
18			Choose B	Choose D			

Option E has no return, so if any options have a positive return, we can eliminate *E*. They do, so eliminate *E*.

Project *C* has an IRR less than MARR. Eliminate *C*.

Determine the incremental IRR of the first set, *B−A*. The investment returns 30.3%, so the higher cost project is justified. Choose *B*.

Next, determine the incremental IRR of the next set, *D−B*. The IRR is 10.0%, which meets the MARR and justifies the higher investment in *D*. Choose *D*.

8-8

An industrial laboratory is comparing options for getting new lab equipment. They can pay cash, finance the purchase, or lease the equipment.

a. Cash price is $35,000; no salvage value; life is 5 years.

b. If financed, price is $34,000; annual interest rate is 10%, payable monthly over 60 months. No salvage value.

c. Lease is $660 per month, payable at the beginning of the month for 60 months. Requires an extra payment of $500 at the beginning of the lease.

Develop a choice table based on EUAC for nominal interest rates from 0% to 20%.

Solution

Calculate the EUAC for each option, based on different interest rates.

	A	B	C	D	E
1	10%	Interest rate			
2	$35,000	Cost of purchase			
3	$34,000	Price if financed			
4	60	Loan period			
5	$660	Lease cost			
6	$500	Upfront lease cost			
7	$743.65	Payment			
8					
9	EUAC				
10	Rate	Purchase	Finance	Lease	Select
11	0%	$583.33	$722.40	$668.33	Purchase
12	2%	$613.47	$722.40	$669.86	Purchase
13	4%	$644.58	$722.40	$671.41	Purchase
14	6%	$676.65	$722.40	$672.97	Lease
15	8%	$709.67	$722.40	$674.54	Lease
16	10%	$743.65	$722.40	$676.12	Lease
17	12%	$778.56	$722.40	$677.72	Lease
18	14%	$814.39	$722.40	$679.33	Lease
19	16%	$851.13	$722.40	$680.96	Lease
20	18%	$888.77	$722.40	$682.60	Lease
21	20%	$927.29	$722.40	$684.25	Lease

The precise breakeven rate between purchase and lease can be calculated using incremental analysis. Find the breakeven interest rate by calculating the RATE of the incremental cash flows.

	Lease	Purchase	Increment	
First cost	-$500	-$35,000	-$34,500	
Operating	-$660	$0	$660	
Rate		Monthly	0.46%	=RATE(nper,pmt,pv)
		Effective	5.7%	=EFFECT(monthly rate*12,12)

Chapter 9

Other Analysis Techniques

FUTURE WORTH

9-1

A new automobile offers free maintenance during the first year of ownership. The maintenance costs the second year will be $100, increasing by $100 each year thereafter. Assume that you will own the automobile for 5 years and that your cost of money is 8%. Find the future worth of the maintenance costs.

Solution

$$FW = 100(P/G, 8\%, 5)(F/P, 8\%, 5) = \$1083$$

9-2

Determine the future worth of 20 quarterly lease payments of $500 at an interest rate of 8%.

Solution

$$FW = [500 + 500(P/A, 2\%, 19)](F/P, 2\%, 20) = \$12,391.75$$

Alternate solution:
$$FW = [500(F/P, 2\%, 1)](F/A, 2\% \ 20) = \$12,391.47$$

9-3

Macoupin Mining Inc. must purchase a new coring machine that costs $60,000 and will last 15 years with a salvage value of $12,000. The annual operating expenses will be $9000 the first year, increasing by $200 each year thereafter. The annual income is $15,000 per year. If Macoupin's MARR is 8%, determine the net future worth of the machine purchase.

Solution

$$NFW = -60,000(F/P, 8\%, 15) - [9000 + 200(A/G, 8\%, 15)](F/A, 8\%, 15)$$
$$+ 15,000(F/A, 8\%, 15) + 12,000$$
$$NFW = \$-45,786$$

9-4

Zap Bug Killers Inc. recently purchased new electrical shock equipment. The equipment cost $16,250 and has a useful life of 4 years. Each year the equipment will produce income of $5500. The costs to operate the equipment are $500 the first year, increasing by $250 a year thereafter. The equipment should have a salvage value of $800. If ZAP's MARR is 8%, what is the net future worth of the equipment? Was the purchase a wise investment?

Solution

$$NFW = -16,250(F/P, 8\%, 4) + [5000 - 250(A/G, 8\%, 4)](F/A, 8\%, 4) + 800$$
$$= -\$351.61 \rightarrow \text{Not a wise investment.}$$

9-5

Salty Nuts Inc. must buy a new nut-shelling machine. The industrial engineer has collected the following information concerning the apparent best alternative. Calculate the net future worth of the alternative if the MARR is 6%.

First cost	$250,000
Annual benefits	73,000 the first year, decreasing by $1200 each year thereafter
Annual O & M costs	28,000 the first year, increasing by $1600 each year thereafter
Salvage value	42,000
Useful life	6 years

Solution

$$NFW = -250,000(F/P, 6\%, 6) + [45,000 - 2800(A/G, 6\%, 6)](F/A, 6\%, 6) + 42,000$$
$$= -\$44,380$$

9-6

An engineer is considering the purchase of a new set of batteries for an electric pallet jack. Given the cost, annual benefit, useful life, and $i = 5\%$, conduct a net future worth analysis to decide which alternative to purchase.

	_A_	_B_
Cost	$19,000	$11,000
Annual benefit	4,000	4,250
Useful life	6 years	3 years

Solution

Alternative _A_:
$$NFW = -19,000(F/P, 5\%, 6) + 4000(F/A, 5\%, 6) = \$1748$$
Alternative _B_:
$$NFW = -11,000(F/P, 5\%, 6) - 11,000(F/P, 5\%, 3) + 4250(F/A, 5\%, 6) = \$1430.50$$
Choose _A_, the larger NFW.

9-7

Lucky Luis has just won $20,000 and wants to invest it for 12 years. There are three plans available to him.

a.   A savings account that pays 3¾% per year, compounded daily.

b.   A money market certificate that pays 6¾% per year, compounded semiannually.

c.   An investment account that, based on past experience, is likely to pay 8½% per year, compounded annually.

If Luis did not withdraw any interest, how much would be in each of the three investment plans at the end of 12 years?

Solution

a. $F = P(1 + i)^n$

$$i_{eff} = \left(1 + \frac{r}{m}\right)^m - 1 \quad = \left(1 + \frac{0.0375}{365}\right)^{365} - 1 = 3.82\%$$

FW $= \$20,000(1 + .0382)^{12} = \$31,361.89$

b. $i_{eff} \left(1 + \frac{0.0675}{2}\right)^2 - 1 = 6.86\%$

FW $= \$20,000(1 + 0.0686)^{12} = \$44,341.67$

c.   FW $= \$20,000(1 + 0.085)^{12} = \$53,233.72$

9-8

The following investment opportunities are available. Use future worth analysis and a MARR of 6% to determine which, if either, alternative should be selected.

	_A_	_B_
First cost	$22,000	$30,000
Annual benefits	6,000	10,000
Annual cost	1,000	3,500
Midlife overhaul	4,000	7,500
Salvage value	3,000	8,000
Useful life	6 years	6 years

Solution

Investment _A_:

NFW $= -22,000(F/P, 6\%, 6) + 5000(F/A, 6\%, 6) + 4000(F/P, 6\%, 3) + 3000 = \$1893$

Investment B:

NFW $= -30,000(F/P, 6\%, 6) + 10,000(F/A, 6\%, 6) + 7500(F/P, 6\%, 3) + 8000 = \$1835$

Choose Investment A, highest NFW.

BENEFIT/COST RATIO

9-9

Rash, Riley, Reed, and Rogers Consulting has a contract to design a major highway project that will provide service from Memphis to Tunica, Mississippi. R^4 has been requested to provide an estimated B/C ratio for the project, summarized as follows.

Initial cost	$25,750,000
Right-of-way maintenance	550,000
Resurfacing (every 8 years)	10% of first cost
Shoulder grading and rework (every 6 years)	1,000,000
Average number of road users per year	1,950,000
Average time savings value per road user	$2

Determine the B/C ratio if $i = 8\%$.

Solution

$AW_{BENEFITS} = 1,950,000 \times \$2 = \$3,900,000$

$AW_{COSTS} = 25,750,000(A/P, 8\%, \infty) + 550,000 + 0.10(25,750,000)(A/F, 8\%, 8)$
$\qquad\quad + 1,000,000(A/F, 8\%, 6) = \$2,988,350$

$B/C = \dfrac{AW_{BENEFITS}}{AW_{COSTS}} = \dfrac{3,900,000}{2,988,350} = 1.31$

9-10

A proposed bridge on the interstate highway system is being considered at the cost of $12 million. It is expected that the bridge will last 20 years. Construction costs will be paid by the federal and state governments. Operation and maintenance costs will be $180,000 per year. Benefits to the public will be $1,500,000 per year. The building of the bridge will result in a cost of $200,000 per year to the general public. The project requires a 6% return. Determine the B/C ratio for the project. State any assumptions made about benefits or costs.

Solution

$200,000 cost to general public is a disbenefit.

$AW_{BENEFITS} = 1,500,000 - 200,000 = \$1,300,000$

$AW_{COSTS} = 12,000,000(A/P, 6\%, 20) + 180,000 = \$1,226,400$

$B/C = \dfrac{AW_{BENEFITS}}{AW_{COSTS}} = \dfrac{1,300,000}{1,226,400} = 1.06$

9-11

The town of Podunk is considering building a new downtown parking lot. The land will cost $25,000, and the construction cost of the lot will be $150,000. Each year, costs associated with the lot will be $17,500. The income from the lot is $18,000 the first year, increasing by $3500 each year for the 12-year expected life of the lot. Determine the B/C ratio if Podunk uses a cost of money of 4%.

Solution

$PW_{BENEFITS} = 18,000(P/A, 4\%, 12) + 3500(P/G, 4\%, 12) = \$334,298$

$PW_{COSTS} = 175,000 + 17,500(P/A, 4\%, 12) = 339,238$

$B/C = \dfrac{AW_{BENEFITS}}{AW_{COSTS}} = \dfrac{334,398}{339,238} = 0.99$

9-12

Tires-R-Us is considering the purchase of new tire-balancing equipment. The machine, which will cost $12,699, will result in annual savings of $1500 with a salvage value at the end of 12 years of $250. For a MARR of 6%, use B/C analysis to determine whether the equipment should be purchased.

Solution

$PW_{BENEFITS} = \$1,500(P/A, 6\%, 12) + \$250(P/F, 6\%, 12) = \$12,700.25$

$PW_{COSTS} = \$12,699$

$B/C = \dfrac{AW_{BENEFITS}}{AW_{COSTS}} = \dfrac{12,700}{12,699} = 1.00$

Conclusion: Yes, the machine should be purchased.

9-13

Dunkin City wants to build a new bypass between two major roads that will cut travel time for commuters. The road will cost $16,000,000 and save 17,500 people $100/year on gas. The road will need to be resurfaced every year at a cost of $10,000. The road is to be used for 20 years. Use B/C analysis to determine whether Dunkin City should build the road. The cost of money is 8%.

Solution

$PW_{COSTS} = 16,000,000 + 10,000(P/A, 8\%, 20) = \$16,098,180$

$PW_{BENEFITS} = (17,500)(100)(P/A, 8\%, 20) = \$17,181,500$

$B/C = \dfrac{AW_{BENEFITS}}{AW_{COSTS}} = \dfrac{17,181,500}{16,098,180} = 1.07$

Conclusion: Yes, Dunkin City should build the bypass.

PAYBACK PERIOD

9-14

For calculating payback period, when is the following formula valid?

$$\text{Payback period} = \frac{\text{First Cost}}{\text{Annual Benefits}}$$

Solution

a. When there is a single cost occurring at time zero (first cost).

b. When Annual Benefits = <u>Net</u> Annual Benefits after any annual costs have been subtracted.

c. When Net Annual Benefits are <u>uniform.</u>

9-15

Is the following statement true or false?

If two investors are considering the same project, the payback period will be longer for the investor with the higher minimum attractive rate of return (MARR).

Solution

Because payback period ignores the MARR, it will be the same for both investors. The statement is false.

9-16

What is the payback period for a project with the following characteristics, given a MARR of 10%?

First cost	$20,000
Annual benefits	8,000
Annual maintenance	2,000 in Year 1, then increasing by $500 per year
Salvage value	2,000
Useful life	10 years

Solution

Payback occurs when the sum of <u>net</u> annual benefits is equal to the first cost. Time value of money is ignored.

Year	Benefits	–	Costs	=	Net Benefits	Total Net Benefits
1	8000	–	2000	=	6000	6,000
2	8000	–	2500	=	5500	11,500
3	8000	–	3000	=	5000	16,500
4	8000	–	3500	=	4500	21,000 > 20,000

Payback period = 3 years + (20,000 – 16,500)/4500 = 3.78 years.

9-17

Determine the payback period (to the nearest year) for the following project if the MARR is 10%.

First cost	$10,000
Annual maintenance	500 in Year 1, increasing by $200 per year
Annual income	3,000
Salvage value	4,000
Useful life	10 years

Solution

Year	Net Income	Sum
1	2500	2500
2	2300	4800
3	2100	6900
4	1900	8800
5	1700	10,500 > 10,000

Payback period = 4 years + (10,000 – 8800)/1700 = 4.71 years.

9-18

Determine the payback period (to the nearest year) for the following project.

Investment cost	$22,000
Annual maintenance costs	1,000
Annual benefits	6,000
Overhaul costs	7,000 every 4 years
Salvage value	2,500
Useful life	12 years
MARR	10%

Solution

Year	Σ Costs	Σ Benefits	
0	22,000	—	
1	23,000	6,000	
2	24,000	12,000	
3	25,000	18,000	
4	33,000	24,000	
5	34,000	30,000	
6	35,000	36,000	← Payback

Payback period = slightly less than 6 years.

More precisely, this will require 5 years plus
$$34,000 + 1000x = 30,000 + 6000x$$
x = 0.8 years   Payback occurs in 5.8 years

9-19

A cannery is considering different modifications to some of their can fillers in two plants that have substantially different types of equipment. These modifications will allow better control and efficiency of the lines. The required investments amount to $135,000 in Plant A and $212,000 for Plant B. The expected benefits (which depend on the number and types of cans to be filled each year) are as follows.

	Plant A	Plant B
Year	Benefits	Benefits
1	$ 73,000	$ 52,000
2	73,000	85,000
3	80,000	135,000
4	80,000	135,000
5	80,000	135,000

a.   Assuming that MARR = 10%, which alternative should be chosen?
b.   Which alternative should be chosen based on payback period?

Solution

a. May be solved in various ways. Using PW

$$PW_A = -135K + 73K(P/A, 10\%, 2) + 80K(P/A, 10\%, 3)(P/F, 10\%, 2)$$
$$= \$156,148.50$$
$$PW_B = -212K + 52K(P/F, 10\%, 1) + 85K(P/F, 10\%, 2) + 135K(P/A, 10\%, 3)(P/F, 10\%, 2)$$
$$= \$182,976.80$$

Therefore, modifications to Plant B are more profitable.

b.

	Plant A			Plant B	
		Cumulative			Cumulative
Year	Benefits	Benefits		Benefits	Benefits
1	73,000	73,000		52,000	52,000
2	73,000	146,000*		85,000	137,000
3	80,000	226,000		135,000	272,000**

*The PBP of A is 1.85 years.   **The PBP of B is 2.55 years.

Based on payback, Plant A has the shortest payback period and should be chosen.

9-20

In this problem the minimum attractive rate of return is 10%. Three proposals are being considered.

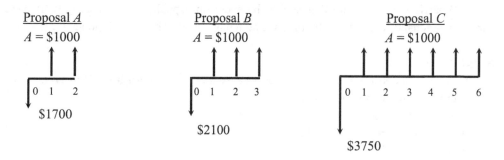

a.   Which proposal would you choose using future value analysis?
b.   How many years for payback for each alternative?

Solution

a.   Proposal A: EUAB = 1000 − 1700(A/P, 10%, 2) = $20.50
     Proposal B: EUAB = 1000 − 2100(A/P, 10%, 3) = $155.60
     Proposal C: EUAB = 1000 − 3750(A/P, 10%, 6) = $139.00

     Proposal A: FW = −1700(F/P,10%,2) + 1000(F/A, 10%, 2) = $ 43
     Proposal B: FW = −2100(F/P,10%,3) + 1000(F/A, 10%, 3) = $ 515
     Proposal C: FW = −3750(F/P,10%,6) + 1000(F/A, 10%, 6) = $1071
     FW recommends Proposal C.

b.   Proposal A: $\dfrac{1700}{1000} = 1.7$ years

      Proposal B $\dfrac{2100}{1000} = 2.1$ years

      Proposal C $\dfrac{3750}{1000} = 3.75$ years

Payback recommends Proposal A.

BREAKEVEN

9-21

A road can be paved with either asphalt or concrete. Concrete costs $20,000/mile and lasts for 20 years. What is the maximum that should be spent on asphalt, which lasts only 10 years? The annual maintenance costs are $500/mile for both pavements. The cost of money = 8%.

Solution

Maintenance doesn't affect the answer because it is the same for both. However, there is nothing wrong with including it.

$20,000(A/P, 8\%, 20) = P_{ASPHALT}(A/P, 8\%, 10)$

$20,000(0.1019) = P_{ASPHALT}(0.1490)$

$P_{ASPHALT} = \$13,678$

9-22

What is the minimum acceptable annual income from a project that has a $70,000 investment cost and a $14,000 salvage value if the life is 15 years and the minimum attractive rate of return (MARR) is 9%?

Solution

$0 = -70,000(A/P, 9\%, 15) + 14,000(A/F, 9\%, 15) + \text{Income}$

$\text{Income} = 70,000(A/P, 9\%, 15) - 14,000(A/F, 9\%, 15) = \$8210$

9-23

Junker Rental Car has a contract with a garage for major repair service for $450 per car, every 6 months. Management estimates that for $350,000, the company could have its own facility, financed at 8% interest for 20 years, and a salvage value of $20,000. They will do their own car repairs at a cost of $200 per car, every 6 months. Ignoring taxes and other economic factors, what is the minimum number of cars needed to make the change feasible?

Solution

Let N = number of autos needed. Note that each time period is 6 months.

$-450N = -350,000(A/P, 4\%, 40) + 20,000(A/F, 4\%, 40) - 200N$

$-250N = -17,465$

$N = 69.86$, or 70 autos needed

9-24

The annual income from an apartment house is $33,600. The annual expense is estimated to be $8000. If the apartment house can be bought today for $349,000, what is the breakeven resale price in 10 years with a 6% interest rate?

Solution

$P = (A_{\text{INCOME}} - A_{\text{EXPENSES}})(P/A, i, n) + F_{\text{RESALE}}(P/F, i, n)$

$349,000 = (33,600 - 8000)(P/A, 6\%, 10) + F_{\text{RESALE}} (P/F, 6\%, 10)$

$349,000 = 25,600(7.360) + F_{\text{RESALE}} (0.5584)$

$F_{\text{RESALE}} = \$287,578.80$

9-25

A machine, costing $16,000 to buy and $1200 per year to operate, will produce savings of $2500 per year for 8 years. If the interest rate is 8%, what is the minimum salvage value that would make the machine an attractive investment?

Solution

$0 = -16,000 + (2500 - 1200)(P/A, 8\%, 8) + SV(P/F, 8\%, 8)$

$0 = -2171.90 + 0.5403SV$

$SV = \$4019.80$

9-26

ABC Manufacturing has a MARR of 12% on new investments. What uniform annual benefit would Investment B have to generate to make it preferable to Investment A?

Year	Investment A	Investment B
0	-$60,000	-$45,000
1-6	+15,000	?

Solution

NPW of A = $-60 + 15(P/A, 12\%, 6) = 1.665$

NPW of $B \geq 1.665 = -45 + B(P/A, 12\%, 6) = -45 + B(4.111)$

NPW of $B \geq (1.665 + 45)/4.111$

$B \geq \$11,351$ per year

9-27

Over the next 6 years, investment in a crane is expected to produce profit from its rental as shown. Assume that the salvage value is zero. Assuming 12% interest, what is the breakeven cost of the crane?

Year	Profit
1	$15,000
2	12,500
3	10,000
4	7,500
5	5,000
6	2,500

Solution

$PW_{PROFIT} = 15,000(P/A, 12\%, 6) - 2500(P/G, 12\%, 6) = \$39,340$

Breakeven Cost = $39,340

9-28

A proposed building may be roofed in either composition roofing (C) or galvanized steel sheet (S). The composition roof costs $56,000 and is replaced every 5 years (assume at the same cost). The steel roof costs $68,000 but the useful life is very long. Neither roof has any salvage value, nor is maintenance needed. If the MARR is 10%, what minimum life must the steel roof have to make it the better alternative? (Report to the nearest whole year; don't bother interpolating.)

Solution

$68,000(A/P, 10\%, n) = 56,000(A/P, 10\%, 5)$

$68,000(A/P, 10\%, n) = 14,772.80 \qquad (A/P, 10\%, n) = 0.2173$

$(A/P, 10\%, 6) = 0.2296$

$(A/P, 10\%, 7) = 0.2054 \qquad\qquad \rightarrow \quad$ So $n = 7$ years

9-29

What is the breakeven cost for Project B if interest equals 10%?

Year	A	B
0	−10,000	?
1–5	+3,500	+2,800

Solution

NPW of $A = -10,000 + 3500(P/A, 10\%, 5) = \$3268.50$

NPW of A = NPW of B

$3268.50 = P_B + 2800(P/A, 10\%, 5)$

$P_B = -\$7346.30$

9-30

The PARC Company can purchase gizmos to be used in building whatsits for $90 each. PARC can manufacture their own gizmos for $7000 per year overhead cost plus $25 direct cost for each gizmo, provided they purchase a gizmo maker for $100,000. PARC expects to use gizmos for 10 years. The gizmo maker should have a salvage value of $20,000 after 10 years. PARC uses 12% as its minimum attractive rate of return. At what annual production rate N should PARC make its own gizmos?

Solution

Equivalent cost solution:

$\text{EAC}_{\text{BUY}} = \$90N$

$\text{EAC}_{\text{MAKE}} = 100,000(A/P, 12\%, 10) + 7000 + 25N - 20,000(A/F, 12\%, 10)$

$\qquad\qquad = 23,560 + 25N$

For breakeven:

$\text{EAC}_{\text{BUY}} = \text{EAC}_{\text{MAKE}}$

$\qquad 90N = 23,560 + 25N$

$N = 362.5$

This indicates that the gizmos should be purchased if 362 or fewer are to be used per year. PARC should make them if 363 or more will be used per year.

9-31

Oliver Douglas decides to install a fuel storage system for his farm that will save him an estimated 6.5 cents/gallon on his fuel cost. Initial cost of the system is $10,000, and the annual maintenance is $25 the first year, increasing by $25 each year thereafter. After a period of 10 years the estimated salvage is $3000. If money is worth 12%, what is the breakeven quantity of fuel?

Solution

$\text{EAC} = 10,000(A/P, 12\%, 10) + 3000(A/F, 12\%, 10) + 25 + 25(A/G, 12\%, 10)$

$\qquad = \$1713.63$

$\text{EAB} = x(0.065) \text{ gallons}$

$\qquad 0 = -1713.63 + x(0.065)$

$x = 26,364 \text{ gallons}$

9-32

Given the following:

$\text{AW}_A = -23,000(A/P, 10\%, 10) + 4000(A/F, 10\%, 10) - 3000 - 3X$

$\text{AW}_B = -8000(A/P, 10\%, 4) - 2000 - 6X$

For these two AW relations, find the breakeven point X in miles per year.

Solution

Setting $AW_A = AW_B$:

$-23,000(A/P, 10\%, 10) + 4000(A/F, 10\%, 10) - 3000 - 3X$

$\quad = -8000(A/P, 10\%, 4) - 2000 - 6X$

$-6491 - 3X = -4524 - 6X$

$X = 655.67$

9-33

To produce an item in-house, equipment costing $250,000 must be purchased. It will have a life of 4 years and an annual cost of $80,000; each unit will cost $40 to manufacture. Buying the item externally will cost $100 per unit. At $i = 12\%$, determine the breakeven production number.

Solution

Buy $= -100X$

Make $= -250,000(A/P, 12\%, 4) - 80,000 - 40X$

$\quad = -82,300 - 80,000 - 40X$

$\quad = -162,300 - 40X$

Setting Buy and Make equal: $-100X = -162,300 - 40X$

$$60X = -162,300$$

$X = 2705$ units

9-34

Data for two drill presses under consideration are listed. Assuming an interest rate of 12%, what salvage value of Press *B* will make the two alternatives equal?

	$\underline{A}$	$\underline{B}$
First cost	$30,000	$36,000
Annual maintenance	1,500	2,000
Salvage value	5,000	?
Useful life	6 years	6 years

Solution

$EAC = P(A/P, i\%, n) - SV(A/F, i\%, n) + \text{Maintenance costs}$

Drill press *A*

$EAC = 30,000(A/P, 12\%, 6) - 5000(A/F, 12\%, 6) + 1500 = \$8180$

Drill press *B*

$EAC = 36,000(A/P, 12\%, 6) - SV(A/F, 12\%, 6) + 2000 = \$10,755.20 - SV(0.1232)$

Setting the two EACs equal:

$8180 = 10,755.20 - SV(0.1232)$ $SV = \$20,903$

9-35

A fruit processing company is considering the purchase of new equipment. The data are as follows.

First cost	$78,750
Annual income	$25/ton of processed fruit
Annual operating costs thereafter	$5500 the first year, increasing $800 each year
Annual property taxes	8% of first cost
Annual insurance	4% of first cost, payable at the beginning of each year
Salvage value	15% of first cost + $1000
Useful life	10 years

The MARR is 4%. Determine the number of tons of fruit that must be processed annually to justify purchasing the machine.

Solution

NPV = 0 at breakeven.

$0 = -78,750 + 25(X)(P/A, 4\%, 10) - [5500(P/A, 4\%, 10) + 800(P/G, 4\%, 10)]$
$\quad - 0.08(78,750)(P/A, 4\%, 10) - [0.04(78,750) + 0.04(78,750)(P/A, 4\%, 9)]$
$\quad + (0.15(78,750) + 1\,A000)(P/F, 4\%, 10)$
$0 = -219,478 + 202.78X$
$X = 1082.35$ tons

9-36

The state Department of Highways is trying to decide whether it should "hot-patch" a short stretch of an existing highway or resurface it. If the hot-patch method is chosen, approximately 300 cubic meters of material would be required at a cost of $600/cubic meter (in place). The shoulders would have to be improved at the same time, at a cost of $24,000. These shoulder improvements must be redone every 2 years (assume the same cost). The annual cost of routine maintenance on the patched road is estimated to be $5000. Alternatively, the state can resurface the road. This surface will last 10 years if maintained properly at a cost of $2000 per year. The shoulders would require reworking at the end of the fifth year at a cost of $5000. Regardless of the method selected, the road will be completely rebuilt in 10 years. At an interest rate of 8%, what is the maximum amount that should be paid for resurfacing the road?

Solution

Hot-Patch

$EUAC = [300(600) + 24,000](A/P, 8\%, 10)$
$\quad + 24,000(A/F, 8\%, 2)(P/A, 8\%, 8)(A/P, 8\%, 10) + 5000 = \$45,277$

Resurface

EUAC = X(A/P, 8%, 10) + 5000(P/F, 8%, 5)(P/A, 8%, 10) + 2000

      = 0.1490X – 2507

X = \$287,047

Appendix 9A

Investing for Retirement and Other Future Needs

9A-1

An initial investment of $1000 in Miracle Plastics has annual returns that have varied a lot. For those returns find the arithmetic average and the geometric mean. Which one is the proper measure?

Year	Annual Return
1	30%
2	-25%
3	-15%
4	35%

Solution

	A	B	C	D	E	F	G
1	Year	Annual Return	1 + Return	Value of Stock Held			
2				1000			
3	1	30%	1.30	1300			
4	2	-25%	0.75	975			
5	3	-15%	0.85	829			
6	4	35%	1.35	1119			
7		6.25%	average				
8			102.85%	geometric mean of 1+return			
9			2.85%	return			
10							
11	*i*	*n*	*PMT*	*PV*	*FV*	Solve for	Answer
12		4	0	-1000	1,119	RATE	2.85%

The geometric mean is the correct measure.

9A-2

A 45-year-old engineer earning $120,000 per year wants to retire at age 65 with $2 million. The engineer has nothing saved and expects to earn 7% annually on the investment?
(a) How much money must be invested each year?
(b) If the employer does a 100% match of retirement savings up to 4% of the employee's salary, how much money must each invest annually?

Solution

	A	B	C	D	E	F	G	H
1		7% Interest rate						
2		20 Years until retirement						
3	$2,000,000	Goal						
4	$120,000	Salary						
5		4% Employer match limit						
6			i	n	PMT	PV	FV	Solve for	Answer
7	(a)	7.00%	20		0	2,000,000	PMT	-$48,786
8	(b)	Employer match						$4,800
9		Employee						-$43,986

9A-3

A 45-year-old engineer earning $110,000 per year wants to retire at age 70 with $1.75 million. The engineer has nothing saved and expects to earn 6% annually on the investment? What fraction of the salary must be invested each year to reach the goal?

Solution

	A	B	C	D	E	F	G	H
1		6% Interest rate						
2		25 Years until retirement						
3	$1,750,000	Goal						
4	$110,000	Salary						
5			i	n	PMT	PV	FV	Solve for	Answer
6			6.00%	25		0	1,750,000	PMT	-$31,897
7								fraction	29.0%

9A-4

A 25-year old engineer wants to spend $40,000 per year traveling as long as possible before switching to saving for retirement. The engineer plans to retire at 60 with $1.5 million and then resume traveling. The engineer expects to earn 8% annually on the investment.
(a) For how many years must the engineer save for retirement?
(b) How long can the engineer travel before beginning to save for retirement?

Solution

	A	B	C	D	E	F	G	H
1		8%	Interest rate					
2	$1,500,000	Goal						
3		$40,000	Annual savings					
4			*i*	*n*	PMT	PV	FV	Solve for	Answer
5	(a)	8.00%			-40,000	0	1,500,000	NPER	18.0
6	(b)		Years until retirement						35.0
7			Years traveling first						17.0

9A-5

An engineering manager retired with investments of $1,200,000 safely invested at 4%. She is 62 and needs $60,000 per year for living expenses, in addition to her social security benefit.

(a) How long will her investment last if it remains invested at 4%?

(b) How much can she spend if it must last until she is 86?

Solution

	A	B	C	D	E	F	G	H
1		4%	Interest rate					
2	$500,000	Savings						
3		-$60,000	(a) Annual cash flow					
4		24	(b) # years of retirement					
5			*i*	*n*	PMT	PV	FV	Solve for	Answer
6	(a)	4.00%			-60,000	500,000		NPER	10.3
7	(b)	4.00%		24		500,000		PMT	-$32,793

9A-6

A new employee puts 10% of his salary of $72,000 into a retirement account. He expects his salary to increase 3% per year. The money is invested in a mutual fund that he expects to average a 6.5% return. How much is in the fund after 10 years?

Solution

	A	B	C	D
1	Salary		72,000	
2	Employee %		10%	
3	Salary increase		3.0%	
4	i		6.5%	
5				FV in 10 years
6	Year	Salary	Deposit	Value
7	1	$72,000	$7,200	$7,200
8	2	$74,160	$7,416	$15,084
9	3	$76,385	$7,638	$23,703
10	4	$78,676	$7,868	$33,111
11	5	$81,037	$8,104	$43,367
12	6	$83,468	$8,347	$54,533
13	7	$85,972	$8,597	$66,675
14	8	$88,551	$8,855	$79,864
15	9	$91,207	$9,121	$94,175
16	10	$93,944	$9,394	$109,691

Chapter 10

Uncertainty in Future Events

10-1

Replacement equipment is being purchased for $40,000, and is expected to reduce costs by 10,000 per year. Estimates of the expected life range from optimistic (12 years) to pessimistic (4 years), with a most likely value of 5 years. There is no salvage value. Determine the optimistic, pessimistic, and most likely rates of return.

Solution

Using a spreadsheet or calculator,

Optimistic: =RATE(12,10000,-40000) = 22.9%

Most Likely: =RATE(5,10000,-40000) = 7.9%

Pessimistic: =RATE(4,10000,-40000) = 0.0%

10-2

Annual savings in labor costs due to an automation project have a most likely value of $35,000. The optimistic value is $45,000 and the pessimistic value is 30,000. What is the expected value for the annual savings?

Solution

$$Expected\ value = \frac{a + 4m + b}{6} = \frac{45,000 + (4)(35,000) + 30,000}{6} = \$35,833$$

10-3

An outdoor construction project is to be completed in the fall. Records show that the chance of sunny weather is 25%, and the chance of early snow is 30%. With sunny weather, the project will cost $22,500. With early snow, the project will cost $30,000. If the weather is moderate, the cost will be $25,000. What is the probability distribution for the construction cost?

Solution

The probability for moderate weather must be calculated: 100% − 25% − 30%, or 45%. This results in the following probability distribution.

State of Nature	Cost	Probability
Sunny	$22,500	0.25
Moderate	25,000	0.45
Early snow	30,000	0.30

10-4

Light rail service between Springfield and Old Saybrook is being considered and will likely cost $900,000 per kilometer. Costs are not certain, and there is a 30% probability that costs will be 40% higher. There is a 20% probability that costs could be 25% lower. What is the probability distribution for the cost per kilometer?

Solution

Most likely cost   = $900,000
Best cost   = 0.75(900,000) = $675,000
Worst cost   = 1.40(900,000) = $1,260,000
The probability for the most likely scenario must be calculated: 100% − 20% − 30%, or 50%.

State of Nature	Cost	Probability
Best	$675,000	0.20
Most Likely	900,000	0.50
Worst	1,260,000	0.30

10-5

Moore Science uses a discount rate of 15% to evaluate projects. Should the following project be undertaken if its life is 10 years and it has no salvage value?

First Cost	P	Net Revenue	P
$200,000	0.1	$70,000	0.3
300,000	0.6	80,000	0.6
500,000	0.3	100,000	0.1

Solution

E(first cost) = 200,000(0.1) + 300,000(0.6) + 500,000(0.3) = $350,000
E(net revenue) = 70,000(0.3) + 80,000(0.6) + 100,000(0.1) = $79,000
E(PW) = −350,000 + 79,000(P/A,15%,10) = $46,500
Yes, do the project.

10-6

Tee-to-Green Golf Inc., is considering the purchase of new automated club assembly equipment. The industrial engineer for TGG thinks that she has determined the "best" choice. However, she is uncertain how to evaluate the equipment because of questions concerning the actual annual savings and salvage value at the end of the expected life. The equipment will cost $500,000 and is expected to last for 8 years. The engineer has the following information concerning the savings and salvage value estimates and the projected probabilities.

	$p = .20$	$p = .50$	$p = .25$	$p = .05$
Savings per year	$65,000	$82,000	$90,000	$105,000
Salvage value	40,000	55,000	65,000	75,000

Determine the NPW if the MARR is 6%.

Solution

E(Savings) = .2(65,000) + .5(82,000) + .25(90,000) + .05(105,000) = $81,750
E(Salvage) = .2(40,000) + .5(55,000) + .25(65,000) + .05(75,000) = $55,500
NPW = −500,000 + 81,750(P/A, 6%, 8) + 55,500(P/F, 6%, 8) = $42,489

10-7

Acme Insurance offers an insurance policy that pays $1000 in reimbursement for luggage lost on a cruise. Historically the company pays this amount in 1 out of every 200 policies it sells. What is the minimum amount Acme must charge for such a policy if the company's goal is to make at least $10 dollars per policy?

Solution

The probability that a loss occurs is $\dfrac{1}{200} = 0.005$

The expected loss to the company is therefore 0.005(1000) = $5

To make a profit of $10 from each policy sold, Acme must charge $15 per policy.

10-8

A roulette wheel consists of 18 black slots, 18 red slots, and 2 green slots. If a $100 bet is placed on black, what is the expected gain or loss? (A bet on black or red pays even money.)

Solution

The probability of black occurring $= \dfrac{18}{38}$

Expected value of the bet $= 100\dfrac{18}{38} - 100\dfrac{20}{38} = \dfrac{100}{19} \approx \$5.26$ loss

10-9

Krispy Kookies is considering the purchase of new dough-mixing equipment. From the estimated NPW and probabilities of the four possible outcomes given, calculate the expected annual worth of the equipment if the life of the equipment is 6 years and $i = 8\%$.

Outcome	NPW	Probability
1	$34,560	.15
2	38,760	.25
3	42,790	.40
4	52,330	.20

Solution

$E(\text{NPW}) = .15(34,560) + .25(38,760) + .40(42,790) + .20(52,330) = \$42,456$

$\text{AW} = 42,456(A/P, 8\%, 6) = \$9183.23$

10-10

The probability that a machine will last a certain number of years is tabulated as follows.

Years of Life	Probability of Obtaining Life
10	.15
11	.20
12	.25
13	.20
14	.15
15	.05

What is the expected life of the machine?

Solution

Expected value $= 10(.15) + 11(.20) + 12(.25) + 13(.20) + 14(.15) + 15(.05) = 12.15$ years

10-11

In the game of craps, two dice are tossed. One of the many bets available is the "Hard-way 8." A $1 bet will return to the bettor $4 if in the game the two dice come up 4 and 4 prior to one of the other ways of totaling eight. For the $1 bet, what is the expected result?

Solution

There are five ways of rolling an 8.→

Die 1	Die 2
2	6
3	5
4	4
5	3
6	2

Hard-way 8 probability = 1/5.

$E(\$) = 1/5(\$4) + 4/5(\$0) = \$0.80$

10-12

Consolidated Edison Power is evaluating the construction of a new electric generation facility. The two choices are a coal-burning plant (CB) and a gaseous diffusion (GD) plant. The CB plant will cost $160 per megawatt to construct, and the GD plant will cost $180 per megawatt. Owing to uncertainties concerning fuel availability and the impact of future regulations related to air and water quality, the useful life of each plant is unknown, but the following probability estimates have been made.

	Probability	
Useful Life (years)	CB Plant	GD Plant
10	.10	.05
20	.50	.25
30	.30	.50
40	.10	.20

a.   Determine the expected life of each plant.
b.   Based on the ratio of construction cost per megawatt to expected life, which plant would you recommend that Con Ed build?

Solution

a.   Expected life
     Coal burning = .10(10) + .50(20) + .30(30) + .10(40) = 24 years
     Gaseous diffusion = .05(10) + .25(20) + .50(30) + .20(40) = 28.5 years

b.   Ratios

Coal burning = 160/24 = $6.66 per megawatt per year
Gaseous diffusion = 180/28.5 = $6.32 per megawatt per year

Recommend the gaseous diffusion plant.

10-13
Crush Cola Company must purchase a bottle-capping machine. The following is known about the machine and about possible cash flows.

	$p = .30$	$p = .50$	$p = .20$
First cost	$40,000	$40,000	$40,000
Annual savings	2,000	3,500	5,000
Annual costs	7,000	5,000	4,000
Actual salvage value	4,000	5,000	6,500

The machine is expected to have a useful life of 10 years. Crush has a MARR of 6%. Determine the NPW of the machine.

Solution

E(Net Saving/Costs) = $(2000 - 7000)(.30) + (3500 - 5000)(.50) + (5000 - 4000(.20)$

$\qquad\qquad = -\$2050$

E(Salvage Value) = $4000(.30) + 5000(.50) + 6500(.20)$

$\qquad\qquad = \$5000$

NPW = $-40,000 - 2050(P/A, 6\%, 10) + 5000(P/F, 6\%, 10) = -\$52,296.00$

10-14
The two finalists in a tennis tournament are playing for the championship. The winner will receive $60,000 and the runner-up $35,000. Determine the expected winnings for each participant if the players are considered to be evenly matched. What would the expected winnings be if one player were favored by 4-to-1 odds?

Solution

Evenly matched, both players' expected winnings will be the same.
Winnings = .5(60,000) + .5(35,000) = $47,500

Assume that Player A is favored by 4-to-1 odds.
The probability that A will win is then 4/5 or 0.8.

Player A's expected winnings = 0.8(60,000) + 0.2(35,000) = $55,000
Player B's expected winnings = 0.2(60,000) + 0.8(35,000) = $40,000

10-15

A new heat exchanger must be installed by CSI Inc. Alternative *A* has an initial cost of $33,400, and Alternative *B* has an initial cost of $47,500. Both alternatives are expected to last 10 years. The annual cost of operating the heat exchanger depends on ambient temperature in the plant and on energy costs. The estimate of the cost and probabilities for each alternative is given. If CSI has a MARR of 8% and uses rate of return analysis for all capital decisions, which exchanger should be purchased?

	Annual Cost	Probability, *p*
Alternative *A*	$4500	.10
	7000	.60
	8000	.25
	9250	.05
Alternative *B*	$4000	.20
	5275	.60
	6450	.15
	8500	.05

Solution

Alternative *A*

E(Annual cost) = .10(4500) + .60(7000) + .25(8000) + .05(9250) = $7112.50

Alternative *B*

E(Annual cost) = .20(4000) + .60(5275) + .15(6450) + .05(8500) = $5357.50

Incremental analysis is required.

$B - A$

NPW = 0 at IRR

$0 = (-47,500 - (-33,400)) + (-5375.50 - (-7112.50))(P/A, i, 10)$

$0 = -14,100 + 1,737(P/A, i, 10)$

$(P/A, i, 10) = 8.12$

$i = 4\%$ $P/A = 8.111;$ IRR ≈ 4%

CSI should purchase the less expensive alternative, Alternative *A*.

10-16

A dam is being considered to reduce river flooding in the Forked River Basin. Information concerning the possible alternatives is given.

Dam Height, H (ft)	First Cost	Annual Probability of Flood if Height =	Damages if Flooding Occurs
0	$ 0	0.25	$800,000
20	700,000	0.05	500,000
30	800,000	0.01	300,000
40	900,000	0.002	200,000

Which dam height minimizes the expected total annual cost? The state uses an interest rate of 5% for flood protection projects, and all dams must last 50 years.

Solution

$H = 0$ (No dam is built.)
EUAC = 800,000(.25) = $200,000
$H = 20$ ft
EUAC = 700,000(A/P, 5%, 50) + 500,000(.05) = $63,360
$H = 30$ ft
EUAC = 800,000(A/P, 5%, 50) + 300,000(.01) = $46,840
$H = 40$ ft
EUAC = 900,000(A/P, 5%, 50) + 200,000(.002) = $49,720
To minimize annual cost, build the 30-foot dam.

10-17

A new product's chief uncertainty is its annual net revenue. Money has been spent, but an additional $30,000 is required to license a patent. The firm's interest rate is 10%. What is the expected PW and standard deviation for deciding whether to proceed?

	State		
	Bad	OK	Great
Probability	0.4	0.5	0.1
Net revenue	−$3,000	$10,000	$25,000
Life & years)	5	5	10

Solution

PW(Bad) = −30,000 + PV(10%, 5, 3000) = −41,372

PW(OK) = −30,000 + PV(10%, 5, -10000) = 7,908

PW(OK) = −30,000 + PV(10%, 10, -25000) = 123,614

E(PW) = 0.4(-41,372) + 0.5(7,908) + 0.1(123,614) = −$233

	State			E(PW)
	Bad	OK	Great	
Probability	0.4	0.5	0.1	
Net Revenue	$−3,000	$10,000	$25,000	
Life (yrs)	5	5	10	
PW	−41,372	7,908	123,614	−233

$$\sigma_{Bad}^2 = (PW^2)(P_{Bad}) - EV(PW)^2 = (-41,372)^2(0.4) - (-233)^2 = 684.6M$$

$$\sigma_{OK}^2 = (PW^2)(P_{OK}) - EV(PW)^2 = (7,908)^2(0.5) - (-233)^2 = 31.2M$$

$$\sigma_{Great}^2 = (PW^2)(P_{Great}) - EV(PW)^2 = (123,614)^2(0.1) - (-233)^2 = 1,528.0M$$

$$\sigma_{PW} = \sqrt{\sigma_{Bad}^2 + \sigma_{OK}^2 + \sigma_{Great}^2} = \sqrt{684.6M + 31.2M + 1,528.0M} = 47,369$$

Appendix 10A

Diversification Reduces Risk

10A-1

An initial investment of $1000 in Hi-Tech Chain has annual returns that have varied a lot. For those returns find the arithmetic average and the geometric mean. Which one is the proper measure?

Year	Annual Return
1	27%
2	-25%
3	30%
4	-33%
5	31%

Solution

	A	B	C	D	E	F	G
1	Year	Annual Return	1 + Return	Value of Stock Held			
2				1000			
3	1	27%	1.27	1270			
4	2	-25%	0.75	953			
5	3	30%	1.30	1238			
6	4	-33%	0.67	830			
7	5	31%	1.31	1087			
8		6.00%	average				
9			101.68%	geometric mean of 1+return			
10			1.68%	return			
11							
12	*i*	*n*	*PMT*	*PV*	*FV*	Solve for	Answer
13		5	0	-1000	1,087	RATE	1.68%

The geometric mean is the correct measure.

138

10A-2

A 35-year-old engineer earning $98,000 per year wants to retire at age 60 with $2.5 million. The engineer has nothing saved and expects to earn 9.6% annually on a 70% stock/30% bond portfolio? How much must be saved each year to reach the goal?

Solution

The goal seems unrealistic since saving more than $25,000 per year is more than a quarter of the pre-tax salary.

	A	B	C	D	E	F	G	H
1	10%	Interest rate						
2	25	Years until retirement						
3	$2,500,000	Goal						
4	$98,000	Salary						
5		i	n	PMT	PV	FV	Solve for	Answer
6		9.60%	25		0	2,500,000	PMT	-$26,992
7							fraction	27.5%

10A-3

A 25-year old engineer wants to save $15,000 per year until retiring at 65. Her plan is that after 5 years she will spend the savings on buying a house and later savings will be for retirement. As her income increases she wants to spend more on herself and her family, which is why she chose a fixed dollar amount to save. Her investing mix will become more conservative over the years so she expects her investment returns to average 5.5% over inflation. How much will she have saved for the house? How much will she have saved at retirement?

Solution

	A	B	C	D	E	F	G	H
1	5.50%	Interest rate						
2	35	Years until retirement						
3	$15,000	Annual savings						
4		i	n	PMT	PV	FV	Solve for	Answer
5	for house	5.50%	5	-$15,000	0		FV	$83,716
6	retirement	5.50%	35	-$15,000	0		FV	$1,503,770

10A-4

An engineer nearing retirement at 66 with $750,000 invested is planning to shift to a portfolio of 20% T-bills, 40% Treasury bonds, and 40% stocks and expects to earn 4% over inflation. How much can the engineer withdraw each year if dying at 100 is the plan?

Solution

	A	B	C	D	E	F	G	H
1	4%	Interest rate						
2	34	Years after retirement						
3	$750,000	Savings						
4			*i*	*n*	*PMT*	*PV*	*FV*	Solve for	Answer
5			4.00%	34		-750,000	0	PMT	$40,736

10A-5

An engineer changed jobs and is signing up for benefits. The company 401(k) includes a low-cost fund that is expected to earn 5.3% annually. The engineer's employer will contribute up to 2% by matching half the employee contribution. So she will save at least 4% of her salary of $80,000 into the account. She expects her salary to increase 2.5% per year. What is the value of the account after 15 years if she deposits the 4% minimum?

Solution

	A	B	C	D
1	80,000	Salary		
2	4%	Employee %		
3	2%	Employer match		
4	2.5%	Salary increase		
5	5.3%	Interest rate		
6				
7	Year	Salary	Deposit	Future Value
8	1	$80,000	$4,800	$4,800
9	2	$82,000	$4,920	$9,974
10	3	$84,050	$5,043	$15,546
11	4	$86,151	$5,169	$21,539
12	5	$88,305	$5,298	$27,979
13	6	$90,513	$5,431	$34,893
14	7	$92,775	$5,567	$42,308
15	8	$95,095	$5,706	$50,256
16	9	$97,472	$5,848	$58,768
17	10	$99,909	$5,995	$67,878
18	11	$102,407	$6,144	$77,620
19	12	$104,967	$6,298	$88,031
20	13	$107,591	$6,455	$99,153
21	14	$110,281	$6,617	$111,024
22	15	$113,038	$6,782	$123,691

10A-6

An engineer changed jobs and is signing up for benefits. The company 401(k) includes a low-cost fund that is expected to earn 5.3% annually. The engineer's employer will contribute up to 2% by matching half the employee contribution. What fraction of her salary (now $80,000) must she save if she wants $250,000 in the account in 15 years? She expects her salary to increase 2.5% per year.

Solution

	A	B	C	D	E
1	$80,000	Salary			
2	10.1%	Employee %			
3	2%	Employer match			
4	2.5%	Salary increase			
5	5.3%	Interest rate			
6	$250,000	Goal in 15 years			
7					
8	Year	Salary	Deposit	Future Value	
9	1	$80,000	$9,702	$9,702	
10	2	$82,000	$9,944	$20,160	
11	3	$84,05	Goal Seek	? ✕	
12	4	$86,15			
13	5	$88,30	Set cell:	D23	
14	6	$90,51	To value:	250,000	
15	7	$92,77	By changing cell:	A2	
16	8	$95,09			
17	9	$97,47	OK	Cancel	
18	10	$99,909	$12,116	$137,192	
19	11	$102,407	$12,419	$156,882	
20	12	$104,967	$12,729	$177,926	
21	13	$107,591	$13,048	$200,404	
22	14	$110,281	$13,374	$224,399	
23	15	$113,038	$13,708	$250,000	

Chapter 11

Depreciation

11-1

Equipment that qualifies for 100% bonus depreciation was purchased for $250,000. Determine the depreciation schedule.

Solution

The cost basis of the asset is $250,000. The depreciation in year 1 is 100%, or $250,000. The book value at the end of year 1 = 250,000 − 250,000 = $0. There is no further depreciation.

11-2

Production equipment has a cost basis of $200,000 and an expected salvage value of $20,000. This equipment qualifies for bonus depreciation. How much bonus depreciation is allowed?
a. in 2020?
b. in 2024?
c. in 2026?

Solution

a. In 2020, 100% bonus depreciation is permitted. 100% × 200,000 = $200,000.
b. In 2024, 60% bonus depreciation is permitted. 60% × 200,000 = $120,000.
c. in 2026, 20% bonus depreciation is permitted. 20% × 200,000 = $40,000.

11-3

A piece of machinery costs $5000 and has an anticipated $1000 resale value at the end of its 5-year useful life. Compute the depreciation schedule for the machinery by the straight-line method.

Solution

$$\text{Annual depreciation charge} = \frac{B - S}{N} = \frac{5000 - 1000}{5} = \$800$$

Depreciation for all 5 years will be $800

11-4

Seed-cleaning equipment was purchased in 2018 for $8500 and was depreciated by the double declining balance (DDB) method for an expected life of 12 years. What is the book value of the equipment at the end of 2023? The original salvage value was estimated to be $2500 at the end of 12 years.

Solution

Year	Depreciation	
2018	$(8500-0)(2/12) =$	$1416.67
2019	$(8500-1416.67)(2/12) =$	1180.56
2020	$(8500-2597.23)(2/12) =$	983.80
2021	$(8500-3581.03)(2/12) =$	819.83
2022	$(8500-4400.86)(2/12) =$	683.19
2023	$(8500-5084.05)(2/12) =$	569.32
	Σ of depreciation	$5653.37

Book value = 8500 – 5653.37 = $2846.63

11-5

Suds-n-Dogs just purchased new automated bun-handling equipment for $12,000. The salvage value of the equipment is anticipated to be $1200 at the end of its 5-year life. Use MACRS to determine the depreciation schedule using a three-year property class.

Solution

Year		Depreciation
1	12,000(.3333)	$3999.60
2	12,000(.4445)	5334.00
3	12,000(.1481)	1777.20
4	12,000(.0741)	889.20

11-6

To meet increased sales, a large dairy is planning to purchase 10 new delivery trucks. Each truck will cost $18,000. Compute the depreciation schedule for each truck, using the modified accelerated cost recovery system (MACRS) method; the recovery period is 5 years.

Solution

Year		Depreciation
1	18,000(.20)	$3600.00
2	18,000(.32)	5760.00
3	18,000(.192)	3456.00
4	18,000(.1152)	2073.60
5	18,000(.1152)	2073.60
6	18,000(.0576)	1036.80

11-7

An asset is purchased for $100,000. The asset is depreciated by using MACRS depreciation and a 5-year recovery period. At the end of the third year of use, the business changes its product mix and disposes of the asset. Determine the depreciation allowed in the third year.

Solution

Disposal before end of MACRS recovery period gives ½-year depreciation in the disposal year.

$D_3 = (0.192/2)(100,000) = \$9600$

11-8

A firm is purchasing office furniture worth $200,000 that has an expected salvage value of $20,000. The furniture is a MACRS 7-year property, but is also eligible for 60% bonus depreciation. Calculate the depreciation schedule.

Solution

Bonus depreciation will be 60% of the investment, or $120,000. MACRS is calculated on the remaining $80,000 balance.

Year	Bonus Depreciation	MACRS	Total
1	$120,000	$11,432	$131,432
2		19,592	19,592
3		13,992	13,992
4		9,992	9,992
5		7,144	7,144
6		7,136	7,136
7		7,144	7,144
8		3,568	3,568
Total	$120,000	$80,000	$200,000

11-9

Computers worth $50,000 are eligible for 20% bonus depreciation plus MACRS. Determine the depreciation schedule.

Solution

Computers are in the 5-year property class. The bonus depreciation will be 20% of the $50,000 investment, or $10,000. The remaining $40,000 will be depreciated using MACRS.

Year	Bonus Depreciation	MACRS	Total
1	$10,000	$8,000	$18,000
2		12,800	12,800
3		7,680	7,680
4		4,608	4,608
5		4,608	4,608
6		2,304	2,304
Total	$10,000	$40,000	$50,000

11-10

An asset will cost $1750 when purchased this year. It is further expected to have a salvage value of $250 at the end of its 5-year depreciable life. Calculate complete depreciation schedules giving the depreciation charge, $D(n)$, and end-of-year book value, $B(n)$, for straight-line (SL), double declining balance (DDB), 100% bonus depreciation, and modified accelerated cost recovery (MACRS) depreciation methods. Assume a MACRS recovery period of 5 years.

Solution

	SL		DDB		100% Bonus		MACRS	
n	$D(n)$	$B(n)$	$D(n)$	$B(n)$	$D(n)$	$B(n)$	$D(n)$	$B(n)$
0		1750		1750		1750		1750.00
1	300	1450	700	1050	1750	0	350.00	1400.00
2	300	1150	420	630	0	0	560.00	840.00
3	300	850	252	378	0	0	336.00	504.00
4	300	550	128	250	0	0	201.60	302.40
5	300	250	0	250	0	0	201.60	100.80
6							100.80	0.00

11-11

Your company is considering the purchase of a secondhand scanning microscope at a cost of $10,500, with an estimated salvage value of $500 and a projected useful life of 4 years. Determine the straight-line (SL), and double declining balance (DDB) depreciation schedules.

Solution

Year	SL	DDB
1	2500	5250
2	2500	2625
3	2500	1312
4	2500	656

11-12

A new machine costs $12,000 and has a $1300 salvage value at the end of its 8-year useful life. Prepare a year-by-year depreciation schedule by the double declining balance (DDB) method.

Solution

DDB depreciation = $\frac{2}{N}$ (B − $\sum$d$_j$)

Year:	1	2	3	4	5	6	7	8*	Total
Depreciation:	3000	2250	1688	1266	949	712	534	301	$10,700
Book value	9000	6750	5062	3796	2847	2135	1601	1300	

*Book value cannot go below declared salvage value. Therefore, the full value of Year 8's depreciation cannot be taken.

11-13

A used piece of depreciable property was bought for $20,000. If it has a useful life of 10 years and a salvage value of $5000, and you use the 150% declining balance schedule, how much will it be depreciated in the 3rd year?

Solution

Depreciation each year = $\frac{1.5}{N}\left(B - \sum d_j\right)$

Depreciation, Yr. 1 = $\frac{1.5}{10}$ (20,000 − 0) = $3,000   BV = 20,000 − 3,000 = 17,000

Depreciation, Yr. 2 = $\frac{1.5}{10}$ (17,000) = $2550   BV = 17,000 − 2,550 = 14,450

Depreciation, Yr. 3 = $\frac{1.5}{10}$ (14,450) = $2168   BV = 14,450 − 2,168 = 12,282

Depreciation will be $2168 during the third year.

11-14

A front-end loader costs $70,000 and has a depreciable salvage value of $10,000 at the end of its 5-year useful life. Use MACRS depreciation to compute the depreciation schedule and book value of the equipment.

Solution

The 5-year recovery period is determined.

Year		Depreciation	Book Value
1	70,000(.20)	$14,000	70,000 – 14,000 = $56,000
2	70,000(.32)	22,400	56,000 – 22,400 = 33,600
3	70,000(.192)	13,440	33,600 – 13,440 = 20,160
4	70,000(.1152)	8,064	20,160 – 8,064 = 12,096
5	70,000(.1152)	8,064	12,096 – 8,064 = 4,032
6	70,000(.0576)	4,032	4,032 – 4,032 = 0

11-15

A pump cost $1000 and has a salvage value of $100 after a life of 5 years. Using the straight-line depreciation method, determine:

a.   The depreciation in the first year.
b.   The book value after 5 years.
c.   The book value after 4 years if the salvage was only $50.

Solution

a. Rate $= \frac{(1000-100)}{5} = \$180$ each year

b. Depreciation after 5 years $= (180)(5) = \$900$. Book value $= \$1000 - 900 = \$100$

c. Rate $= \frac{(1000-50)}{5} = \$190$ each year. Book value $= 1000 - (4)(190) = \$240$

11-16

Nuts-R-Us Inc. purchased nut-shelling equipment at a total cost of $80,000. The equipment was depreciated by using MACRS with a recovery class of 3 years and an anticipated end-of-useful-life value of $8000. The company has decided the equipment is no longer needed after two years and wishes to determine the minimum value it can accept for the equipment (that is, the lowest value that will result in no loss on the sale). Find the minimum selling price for the equipment.

Solution

Disposal before end of MACRS recovery period gives ½-year depreciation in disposal year.
Total depreciation $= 0.3333(80,000) + (0.4445/2)(80,000) = 26,664 + 17,780 = \$44,444$
$BV_2 = 80,000 - 44,444 = \$35,556$

11-17

Thick Trunk Sawmill purchases a new automated log planer for $95,000. The asset is depreciated by using straight-line depreciation over a useful life of 10 years to a salvage value of $5000. Find the book value at the end of Year 6.

Solution

$d_t = (95,000 - 5000)/10 = \$9000/\text{year}$

$\sum_{\text{Depreciation}} = 9000 \times 6 = \$54,000$ after 6 years

$BV_6 = 95,000 - 54,000 = \$41,000$

11-18

Adventure Airlines recently purchased a new baggage crusher for $50,000. It is in a MACRS 7-year property class with estimated salvage value of $8000. Use 40% bonus depreciation with MACRS to determine the depreciation charge on the crusher for the third year of its life and the book value at the end of 3 years.

Solution

Bonus depreciation in year 1 will be worth $(0.40)(50,000) = \$20,000$

The remaining depreciation will be determined using MACRS, using a cost basis of $30,000.

Year	Bonus Depreciation	MACRS	Book Value
1	$20,000	$4287	$25,713
2	0	7347	18,366
3	0	5247	13,119

The book value at the end of 3 years is $13,119.

11-19

Hoppy Hops Inc., purchased hop-harvesting machinery for $150,000 four years ago. Owing to a change in the method of harvesting, the machine was recently sold for $37,500. Determine the MACRS deprecation schedule for the machinery for the 4 years of ownership. Assume a 5-year property class. What is the recaptured depreciation or loss on the sale of the machinery?

Solution

Year	MACRS	Depreciation
1	.2(150,000) =	$30,000
2	.32(150,000) =	48,000
3	.1920(150,000) =	28,800
4	(.1152/2)(150,000) =	8,640

Total depreciation = 30,000 + 48,000 + 28,800 + 8640 = $115,440

Book value at end of Year 4 = 150,000 - 115,440 = $34,560

Recaptured depreciation = 37,500 - 34,560 = $2940

11-20

Equipment costing $100,000 was bought in early 2019 and sold three years later for $20,000. Determine the depreciation recapture, ordinary losses, or capital gains associated with selling the equipment. Consider two cases:

a. 100% bonus depreciation

b. 5-year MACRS

Solution

    a. With 100% bonus depreciation, the equipment is fully depreciated in Year 1, leaving a book value of $0. The equipment is sold for $20,000, so the depreciation recapture will be $20,000.

    b. Book value $= 100,000 - 100,000[0.2000 + 0.3200 + (0.1920/2)] = \$38,400$
    The equipment is sold at a loss of $(38,400 - 20,000) = \$18,400$

11-21

A lumber company purchased a tract of timber for $70,000. The value of the 25,000 trees on the tract was established to be $50,000. The value of the land was established to be $20,000. In the first year of operation, the lumber company cut down 5000 trees. What was the depletion allowance for the year?

Solution

    For standing timber, only cost depletion (not percentage depletion) is permissible. Five thousand of the trees were harvested; therefore $5,000/25,000 = 0.20$ of the tract was depleted. Land cannot be depleted. Only the timber, which is valued at a total of $50,000, is subject to depletion.

    Therefore, the first year's depletion allowance would be $= 0.20(\$50,000) = \$10,000$.

11-22

In the production of beer, a final filtration is accomplished by the use of kieselguhr, or diatomaceous earth, which is composed of the fossil remains of minute aquatic algae a few microns in diameter, and pure silica. A company has purchased a property for $840,000 that contains an estimated 60,000 tons of kieselguhr. Compute the depreciation charges for the first 3 years, given that production (or extraction) of 3000 tons, 5000 tons, and 6000 tons is planned for Years 1, 2, and 3, respectively. Use the cost-depletion method, assuming no salvage value for the property.

Solution

Total diatomaceous earth in property = 60,000 tons

Cost of property = $840,000

Then, $\dfrac{\text{depletion allowance}}{\text{tons extracted}} = \dfrac{\$840,000}{60,000 \text{ tons}} = \$14/\text{ton}$

Year	Tons Extracted	Depreciation Charge
1	3000	3000 × 14 = $42,000
2	4000	4000 × 14 = 56,000
3	5000	5000 × 14 = 70,000

Chapter 12

Income Taxes
for Corporations

12-1

Determine the income taxes for a firm with the following results:

Income from sales	$10.0 million
Total expenses	$ 2.4 million
Depreciation	$ 1.6 million

Solution

Taxable Income = Gross income – expenses – depreciation

Taxable Income = 10 million – 2.4 million – 1.6 million = $6.0 million

Taxes = (0.21)(6.0 million) = $1.26 million

12-2

A small firm has a taxable income of $560,000. They operate in a state that has a corporate income tax rate of 6.5%. What is the total federal and state tax that they must pay? What is their total incremental tax rate?

Solution

State tax = (0.065)(560,000) = $36,400

Federal tax = (0.21)(560,000 – 36,400) = $109,956

Total State + Federal tax = $36,400 + 109,956 = $146,356

Because both state and federal

are flat taxes, the current rate will be the incremental rate, or

146,356 / 560,000 = 0.261 = 26.1%

12-3

A company, whose earnings put it in the 25% marginal tax bracket, is considering the purchase of a new piece of equipment for $25,000. The equipment will be depreciated by using the straight-line method over a 4-year depreciable life to a salvage value of $5000. It is estimated that the equipment will increase the company's earnings by $8000 for each of the 4 years it is used. Should the equipment be purchased? Use an interest rate of 10%.

Solution

Year	BTCF	Depreciation	TI	Taxes	ATCF
0	–25,000				–25,000
1	8,000	5000	3000	750	7,250
2	8,000	5000	3000	750	7,250
3	8,000	5000	3000	750	7,250
4	8,000	5000	3000	750	7,250
	5,000				5,000

SL depreciation = ¼(25,000 – 5000) = $5000
NPV = –25,000 + 7250(*P*/*A*, 10%, 4) + 5000(*P*/*F*, 10%, 4) = $1398
Therefore, purchase the equipment.

12-4

A corporation expects to receive $32,000 each year for 15 years if a particular project is undertaken. There will be an initial investment of $150,000. The expenses associated with the project are expected to be $7500 per year. Assume straight-line depreciation, a 15-year useful life, and no salvage value. Use a combined state and federal 28% marginal tax rate, and determine the project's after-tax rate of return.

Solution

Year	BTCF	Depreciation	TI	Taxes	ATCF
0	–150,000				–150,000
1–15	24,500	10,000	14,470	4060	20,440

	A	B	C	D	E	F	G	H	I
1	Problem	*i*	N	PMT	PV	FV	Solve for	Answer	Formula
2	12-4		12	20,440	-$150,000	0	RATE	8.5%	=RATE(C2,D2,E2,F2)

The after-tax rate of return will be 8.5%.

12-5

A corporation's marginal tax rate is 28%. An outlay of $35,000 is being considered for a new asset. Estimated annual receipts are $20,000 and annual disbursements $10,000. The useful life of the asset is 5 years, and it has no salvage value.

a. What is the prospective rate of return before income tax?

b. What is the prospective rate of return after taxes, assuming straight-line depreciation?

Solution

SL depreciation $= (B - S)/N = \frac{35{,}000 - 0}{5} = \$7000/\text{year}.$

Year	BTCF	Depreciation	TI	Taxes	ATCF
0	−35,000	—	—	—	−35,000
1	10,000	7000	3000	840	9,160
2	10,000	7000	3000	840	9,160
3	10,000	7000	3000	840	9,160
4	10,000	7000	3000	840	9,160
5	10,000	7000	3000	840	9,160

	A	B	C	D	E	F	G	H	I
1	Problem	i	N	PMT	PV	FV	Solve for	Answer	Formula
2	12-5								
3	Before Tax		5	10,000	-35000	0	RATE	13.2%	=RATE(C3,D3,E3,F3)
4									
5	After Tax		5	9,160	-35000	0	RATE	9.7%	=RATE(C5,D5,E5,F5)

    a. The Before-tax ROR is 13.2%

    b. The After-tax ROR is 9.7%.

12-6

A project under consideration by PHI Inc., is summarized. The company uses straight-line depreciation, pays taxes at the 30% marginal rate, and requires an after-tax MARR of 12%. Use net present worth to determine whether the project should be undertaken.

First cost	$75,000
Annual revenues	26,000
Annual costs	13,500
Salvage value	15,000
Useful life	10 years

Solution

Year	BTCF	Depreciation	TI	Taxes	ATCF
0	−75,000	—	—	—	−75,000
1–10	12,500	7500	5000	1500	11,000
10	15,000	—	—	—	15,000

NPW $= -75{,}000 + 11{,}000(P/A, 12\%, 10) + 15{,}000(P/F, 12\%, 10) = -\$8020$

No, do not do the project, since NPW < 0.

12-7

PARC, a large profitable firm, has an opportunity to expand one of its production facilities at a cost of $375,000. The equipment is expected to have an economic life of 10 years and to have a resale value of $25,000 after 10 years of use. If the expansion is undertaken, PARC expects that income will increase by $60,000 for Year 1, and then increase by $5000 each year through Year 10. The annual operating cost is expected to be $5000 for the first year and to increase by $250 per year thereafter. If the equipment is purchased, PARC will depreciate it by using the straight-line method to a zero salvage value at the end of Year 8 for tax purposes. The applicable marginal tax rate is 28%.

If PARC's minimum attractive rate of return (MARR) is 15%, should the firm undertake this expansion?

Solution

	A	B	C	D	E	F	G	H
1	Year	Income	Costs	BTCF	Depreciation	TI	Taxes	ATCF
2	0	0	0	-375,000	0	0	0	-375,000
3	1	60,000	5000	55,000	46,875	8,125	2,275	52,725
4	2	65,000	5250	59,750	46,875	12,875	3,605	56,145
5	3	70,000	5500	64,500	46,875	17,625	4,935	59,565
6	4	75,000	5750	69,250	46,875	22,375	6,265	62,985
7	5	80,000	6000	74,000	46,875	27,125	7,595	66,405
8	6	85,000	6250	78,750	46,875	31,875	8,925	69,825
9	7	90,000	6500	83,500	46,875	36,625	10,255	73,245
10	8	95,000	6750	88,250	46,875	41,375	11,585	76,665
11	9	100,000	7000	93,000	0	93,000	26,040	66,960
12	10	105,000	7250	122,750	0	122,750	34,370	88,380
13	MARR	15%						
14	NPV			-12,136				-54,842

The project should not be undertaken because it has a negative NPV.

12-8

The Salsaz-Hot manufacturing company must replace a machine used to crush tomatoes for its salsa. The industrial engineer favors a machine called the Crusher. Information concerning the machine is given.

First cost	$95,000
Annual productivity savings	19,000
Annual operating costs	6,000
Annual insurance cost*	1,750

Property taxes equal to 5% of the first cost are payable at the end of each year.

*Payable at the beginning of each year

Depreciable salvage value	$10,000
Actual salvage value	14,000
Depreciable life	6 years
Actual useful life	10 years
Depreciation method	SL

Relevant financial information for Salsaz-Hot:

Marginal tax rate	26%
MARR	10%

Determine the net present worth.

Solution

Cash expenses are multiplied by (1 – tax rate).
  Incomes are multiplied by (1 – tax rate).
  Depreciation is multiplied by tax rate.
  Capital recovery is not taxable; therefore it is multiplied by 1.
Recaptured depreciation is multiplied by (1 – tax rate).

Year		
0	First cost	−95,000
1–10	Net savings: 13,000(P/A, 10%, 10)(0.66)	59,115
1–10	Property taxes: 0.05(95,000)(P/A, 10%, 10)(0.66)	−21,600
0–9	Insurance: [1750 + 1750(P/A, 10%, 9)](0.66)	−8,753
1–6	Depreciation: 14,167(P/A, 10%, 6)(0.34)	16,041
10	Capital recovery: 10,000(P/F, 10%, 10)	3,855
10	Recaptured depr.: 4000(P/F, 10%, 10)(0.66)	1,141
	NPW	−$45,201

12-9

A company bought an asset at the beginning of 2018 for $100,000. The company now has an offer to sell the asset for $60,000 at the end of 2019. Using double declining balance, determine the capital loss or recaptured depreciation that would be realized for 2018.

Depreciable Life (years)	Salvage Value*	Recaptured Depreciation	Capital Loss
6	0		

*Assumed for depreciation purposes.

Solution

Depreciable Life (years)	Salvage Value	Recaptured Depreciation	Capital Loss
6	0	15,555	

$$\text{Depreciation Year 1} = \frac{2}{6}(100,000) = 33,333$$

$$\text{Depreciation Year 2} = \frac{2}{6}(100,000 - 33,333) = 22,222$$

$$\text{Book value} = 100,000 - 55,555 = 44,445$$

$$\text{Recaptured depreciation} = 60,000 - 44,445 = \$15,555$$

12-10

Scallop Corporation purchased oil exploration equipment for $600,000 that will be depreciated over 10 years using the double declining balance method. Combined state and federal tax rate is 24%. The equipment may be rented each year for $330,000, and will then be sold after 5 years for $200,000. What is the after-tax rate of return?

Solution

	A	B	C	D	E	F	G
1	Year	Before-Tax Cash Flow	Depr.	Taxable Income	Income Taxes 24%	After-Tax Cash Flow	ATCF
2	0	-$600,000			$0	-$600,000	-$600,000
3	1	330,000	120,000	210,000	-50,400	$279,600	$279,600
4	2	330,000	96,000	234,000	-56,160	$273,840	$273,840
5	3	330,000	76,800	253,200	-60,768	$269,232	$269,232
6	4	330,000	61,440	268,560	-64,454	$265,546	$265,546
7	5	330,000	49,152	280,848	-67,404	$262,596	$461,782
8	5	200,000		3,392	-814	$199,186	
9		Sum	$403,392			IRR	38.6%

The after-tax rate of return will be 38.6%.

12-11

New equipment costing $30,000 has a 5-year life and no salvage value. Benefits are expected to be $8000 per year. The equipment qualifies for 100% bonus depreciation. The firm has a 28% combined marginal income tax rate. What is the after-tax rate of return?

Solution

	A	B	C	D	E	F
1	Year	Before-Tax Cash Flow	Depr.	Taxable Income	Income Taxes	After-Tax Cash Flow
2	0	-$30,000			$0	-$30,000
3	1	8000	30,000	-22,000	-6,160	14,160
4	2	8000		8,000	2,240	5,760
5	3	8000		8,000	2,240	5,760
6	4	8000		8,000	2,240	5,760
7	5	8000		8,000	2,240	5,760
8						
9		Sum	$60,000		IRR	9.2%

The after-tax rate of return is 9.2%.

12-12

Macoupin Mining bought $60,000 of equipment that qualified for 40% bonus depreciation, with the balance using 5-year MACRS depreciation. The equipment saved $16,000 each year, and was sold for $20,000 after 5 years of use. The combined incremental tax rate is 26%, and the firm's after-tax MARR is 12%. Was this a wise investment?

Solution

	A	B	C	D	E	F
1	Year	BTCF	Depr.	Taxable Income	Income Taxes 26%	ATCF
2	0	-$60,000				-$60,000
3	1	16,000	31,200	-15,200	3,952	19,952
4	2	16,000	11,520	4,480	-1,165	14,835
5	3	16,000	6,912	9,088	-2,363	13,637
6	4	16,000	4,147	11,853	-3,082	12,918
7	5	16,000	2,074	13,926	-3,621	28,257
8	5	20,000		15,853	-4,122	
9	Sum		$55,853		IRR	14.3%

Note: There is only ½ year of depreciation in the year of disposal
Gain in year of disposal = 20,000 – (60,000 – 55,853) = $15,853
After-tax rate of return = 14.3%. This was a good investment.

12-13

An asset with 5-year MACRS life will be purchased for $10,000. It will produce net annual benefits of $2000 per year for 6 years, after which time it will have a net salvage value of zero and will be retired. The company's marginal tax rate is 24%. Calculate the after-tax cash flows.

Solution

Year	BTCF	Depreciation	TI	Taxes	ATCF
0	−10,000				−10,000
1	2,000	2000	0	0	2,000
2	2,000	3200	−1200	288	2,288
3	2,000	1920	80	-19	1,981
4	2,000	1152	848	-204	1,796
5	2,000	1152	848	-204	1,796
6	2,000	576	1424	-342	1,658

12-14

A large and profitable company, in the 34% marginal tax bracket, is considering the purchase of a new piece of machinery that will yield benefits of $10,000 for Year 1, $15,000 for Year 2, $20,000 for Year 3, $20,000 for Year 4, and $20,000 for Year 5. The machinery is to be depreciated by using the modified accelerated cost recovery system (MACRS) with a 3-year recovery period. The MACRS percentages are 33.33, 44.45, 14.81, 8.41, respectively, for Years 1, 2, 3, and 4. The company believes the machinery can be sold at the end of 5 years of use for 25% of the original purchase price. If the company requires a 12% after-tax rate of return, what is the maximum purchase cost it can pay?

Solution

Year	BTCF	Depreciation	TI	Taxes	ATCF
0	−P				−P
1	10,000	.3333P	10,000 − .3333P	3400 + .1133P	6,600 + .1133P
2	15,000	.4445P	15,000 − .4445P	5100 + .1511P	9,900 + .1511P
3	20,000	.1481P	20,000 − .1481P	6800 + .0504P	13,200 + .0504P
4	20,000	.0841P	20,000 − .0841P	6800 + .0286P	13,200 + .0286P
5	20,000		20,000	6,800	13,200
	.25P		.25P	.085P	.165P

$$P = 6600(P/A\ 12\%, 3) + 3300(P/G, 12\%, 3) + 13{,}200(P/A, 12\%, 2)(P/F, 12\%, 3)$$
$$+ .1133P(P/F, 12\%, 1) + .1511P(P/F, 12\%, 2) + .0504P(P/F, 12\%, 3)$$
$$+ .0285P(P/F, 12\%, 4) + .165P(P/F, 12\%, 5)$$
$$P = \$61{,}926.52$$

12-15

An office is looking at enhancing their ERP system to replace 5 people. Each person earns $40,000 per year, with benefits counting for an additional 40% of salaries. The office uses 10% as an after-tax interest rate. Maintenance is assumed to be 5% of cost, and property taxes are 2% of cost. Depreciation is straight-line over 5 years, and income taxes are 24%. How much of a computer investment can be justified?

Solution

The known cash flows may be set up in a table, as shown.

One way to solve the problem is using the final column to determine the present value, and using Goal Seek to find a value of P that makes the present value equal to 0. This will be the minimum value of P.

	A	B	C	D	E	F	G
1	Year	Before-Tax Cash Flow	Depr.	Taxable Income	Income Taxes at 24%	After-Tax Cash Flow	Goal Seek
2	0	-P				-P	-791,674
3	1	+280,000 - .07P	0.2*P	+280,000 - .27*P	-67,200 +.065*P	212800 - .005*P	208842
4	2	+280,000 - .07P	0.2*P	+280,000 - .27*P	-67,200 +.065*P	212800 - .005*P	208842
5	3	+280,000 - .07P	0.2*P	+280,000 - .27*P	-67,200 +.065*P	212800 - .005*P	208842
6	4	+280,000 - .07P	0.2*P	+280,000 - .27*P	-67,200 +.065*P	212800 - .005*P	208842
7	5	+280,000 - .07P	0.2*P	+280,000 - .27*P	-67,200 +.065*P	212800 - .005*P	208842
8	i= 10%						
9						PV	0.0

Goal Seek identifies the value of P to be $791,674.

Alternatively,

$P = (212{,}800 - 0.005P)(P/A, 10\%, 5) = (212{,}800 - 0.005P)(3.791) = 806{,}725 - 0.019P$

$1.019P = 806{,}725$

$P = \$791{,}683$

2-16

A developer is deciding whether to purchase some heavy-duty landscaping equipment. The alternative is to rent the equipment. One set of equipment will cost $40,000 with no salvage value, depreciated using straight-line over 5 years. Operating and maintenance costs are $320 per day. Similar equipment can be rented for $600 per day (weekends included). The analysis needs to be on an after-tax basis. The combined marginal income tax rate is 26%, and the after-tax MARR is 10%. How many days per year must the equipment be used in order to justify its purchase?

Solution

The known cash flows may be set up in a table, as shown.

One way to solve the problem is the after-tax cash flow column to determine the NPV, and using Goal Seek to find a value of x (number of days) that makes the NPV equal to 0. This will be the minimum number of days required to justify the purchase.

⬜	A	B	C	D	E	F
1	Year	Before-Tax Cash Flow	Depr.	Taxable Income	Income Taxes 26%	After-Tax Cash Flow
2	0	-$40,000				-$40,000
3	1	280x	8000	280x-8000	72.8x -2080	207.2x + 2080
4	2	280x	8000	280x-8000	72.8x -2080	207.2x + 2080
5	3	280x	8000	280x-8000	72.8x -2080	207.2x + 2080
6	4	280x	8000	280x-8000	72.8x -2080	207.2x + 2080
7	5	280x	8000	280x-8000	72.8x -2080	207.2x + 2080
8	$i =$	10%				
9	$x =$	40.89	Identified using Goal Seek			
10	NPV	0.00	=-F2+PV(B8,A7,207.2*B9+2080,0)			

The number of days required is 41 in order to justify the purchase of the equipment. Alternatively, factors may be used.

$40,000 = 207.2x + 2080)(P/A, 10\%, 5) = 207.2x + 2080)(3.791) = 785.5x + 7885$

$x = (40,000 - 7,885)/785.5 = 40.88$ days   or 41 days

12-17

ACME Coyote Products bought processing equipment for $275,000, eligible for the Section 179 depreciation. The equipment saves the firm $70,000 per year and is good for 5 years, when it will have a negligible salvage value. ACME has a 24% tax rate. What is the after-tax rate of return on this purchase?

Solution

	A	B	C	D	E	F	G
1	Year	Before-Tax Cash Flow	Depr.	Sect.	Taxable Income	Income Taxes	After-Tax Cash Flow
2				179			
3	0	-$275,000				0	-$275,000
4	1	70,000	0	$275,000	-$205,000	$49,200	119,200
5	2	70,000	0		70,000	-16,800	53200
6	3	70,000	0		70,000	-16,800	53200
7	4	70,000	0		70,000	-16,800	53200
8	5	70,000	0		70,000	-16,800	53200
9						IRR	7.8%

The after-tax rate of return is 7.8%.

12-18

A company has purchased a major piece of equipment that has a useful life of 20 years. An analyst trying to decide on a maintenance program has narrowed the choices to two alternatives. Alternative A is to perform $1000 of major maintenance every year. Alternative B is to perform $5000 of major maintenance only every fourth year. In either case, maintenance will be performed during the last year so that the equipment can be sold for $10,000. If the MARR is 12%, which maintenance plan should be chosen? Is it possible that the decision would change if income taxes were considered? Why or why not?

Solution

Equivalent Annual Cost$_A$ = $1000
Equivalent Annual Cost$_B$ = $5000(A/F, 12%, 4) = $1046
Therefore, choose Alternative A.
The decision would not change if taxes were considered. Since the cash flows for both alternatives would be reduced by the same percentage because of taxes, $EAC_A > EAC_B$ would still be true. If we assume a 25% tax rate, for example, the computations are as follows.

Alternative A	Year	BTCF	TI	Taxes	ATCF
	1–4	−1000	−1000	250	−750

EAC$_A$ = $750

Alternative B	Year	BTCF	TI	Taxes	ATCF
	1–3	0	0	0	0
	4	−5000	−5000	1250	−3750

EAC$_B$ = 3750(A/F, 12%, 4) = $785

12-19

A large company must build a bridge to have access to land for expansion of its manufacturing plant. The bridge could be fabricated of normal steel for an initial cost of $30,000 and should last for 15 years. Maintenance will cost $1000 per year. If the steel used were more corrosion resistant, the annual maintenance cost would be only $100 per year, although the life would be the same. In 15 years there would be no salvage value for either bridge. The company pays combined federal and state taxes at the 32% marginal rate and uses straight-line depreciation. If the minimum attractive after-tax rate of return is 12%, what is the maximum amount that should be spent on the corrosion-resistant bridge?

Solution

Steel

Year	BTCF	Depreciation	TI	Taxes	ATCF
0	−30,000				−30,000
1–15	−1,000	2000	−3000	+960	-40

Corrosion-Resistant Steel

Year	BTCF	Depreciation	TI	Taxes	ATCF
0	−P				−P
1–15	−100	P/15	−100 − P/15	(+32 + .021P)	−68 + .021P

$NPW_A = NPW_B$ for breakeven:

−40(P/A, 12%, 15) − 30,000 = (−68 + .021P)(P/A, 12%, 15) − P

−40(6.811) − 30,000 = (−68 + .021P)(6.811) − P

−30,272 = −463 + .143P − P

P = $34,783

Appendix 12A

Taxes and Personal Financial Decision Making

12A-1

A young entry-level engineer, Kelly Green, earned $43,000 last year. Kelly is single and will not itemize his tax return. How much federal income tax does Kelly owe? Assume no other income adjustments.

Solution

Taxable income = Adjusted gross income − standard deduction

$$= 43,000 - 12,000 = 31,000$$

Taxes owed = 952.50 + 12% of the amount over $9525

$$= 952.50 + (0.12)(31,000 - 9525) = \$3529.50$$

12A-2

Tammie pays federal income taxes at the incremental rate of 22% and state income taxes at the incremental rate of 3.4%. What is her combined incremental income tax rate?

Solution

Combined incremental tax rate $= \Delta$ State tax rate $+ [(\Delta$ Federal tax rate$)(1 - \Delta$ State tax rate$)]$

$$= 0.034 + (0.22)(1 - .034)$$

$$= 0.2465 = 24.65\%$$

12A-3

As a senior engineer for a large consulting firm, Ray earned $113,000 last year. Ray is married with two children. His wife had no income and they will not itemize their tax return. What is the effective federal tax rate the couple will pay if they file a joint return? Assume no other income adjustments.

Solution

Taxable income = Adjusted gross income − standard deduction

$$= 113,000 - 24,000 = 89,000$$

Taxes owed = 8,907.00 plus 22% of the amount over $77,400

$$= 8,907 + (0.22)(89,000 - 77,400) = \$11,459.00$$

Effective tax rate $= \dfrac{Taxes\ Paid}{Total\ Income} = \dfrac{11,459}{113,000} = 0.1014 = 10.14\%$

12A-4

Lexie earned $29,995 last year while also attending the University of New London as a graduate student. She paid tuition of $4100 during the year and had additional educational expenses of $900. Lexie is single and will not itemize her tax return. Assuming no other income adjustments, how much federal tax will Lexie owe?

Solution

Taxable income = Adjusted gross income – standard deduction
$$= 29,995 - 12,000 = 17,995$$
Taxes owed = 952.50 + 12% of the amount over $9525
$$= 952.50 + (0.12)(17,995 - 9525) = \$1968.90$$

The Education Expenses tax deduction could be taken. This will deduct $4000 from her income, and reduce her taxes by 12% or $(0.12)(4000) = \$480$.

Based on the amount of educational expenses ($4100), it is to Lexie's advantage to use the American Opportunity Tax Credit, which can reduce taxes up to $2500.

American Opportunity Tax Credit = $2500
Taxes owed = $1968.90 - 2500 = -\$531.10$

Tax owed is –$531.10, which is less than 0, so the tax owed is $0.

12A-5

A student pays her own tuition and fees. Her income is $45,000 and her education expenses are $5200. Using the American Opportunity Tax Credit, what is the maximum tax credit she is due?

Solution

The full American Opportunity Tax Credit is available for people earning less than $80,000, and covers all expenses up to $2000 plus 25% of the next $2000. Therefore, her tax credit will be

$$2000 + 0.25(2000) = \$2500 \qquad \text{which is the maximum allowed.}$$

12A-6

A 1-year savings certificate that pays 15% is purchased for $10,000. If the purchaser pays taxes at the 27% incremental income tax rate, the after-tax rate of return on this investment is

Solution

After-tax ROR = (1 – Tax rate)(Before-tax IRR) = $(1 - 0.27)(0.15) = 10.95\%$

12A-7

A newly hired engineer wants to take advantage of the firm's generous 401k plan, where the company will do a 1:1 match up to 3% of the person's salary. The new engineer is making $60,000 per year. If the engineer is paid twice per month, how much is the person putting into the 401k each pay period? How much is being added each pay period in total?

Solution

The engineer will have 3% of their half-month salary placed in the 401K.

$(60,000/24)(0.03) = \$75.00$

The employer will match this, adding another $75.00 for a total of $150 each pay period.

12A-8

A whole life policy worth $100,000 can be purchased for $1200 per year, payable at the beginning of the year. The insurance company uses 12% of this to fund insurance, and the remaining is put into a savings plan that guarantees a 1.9% rate of return. What is the cash value of the plan after 10 years, when there is no penalty for cashing the policy?

Solution

88% of the $1200 payment (paid at the beginning of the year) will be saved at $i = 1.9\%$ for 10 years.

	A	B	C	D	E	F	G	H	I
1	Problem	*i*	*N*	*PMT*	*PV*	*FV*	Solve for	Answer	Formula
2	12A-8	1.90%	10	-1056.00	$0		FV	$11,510	=FV(B2,C2,D2,E2)

Chapter 13

Replacement Analysis

13-1

An engineer is trying to determine the economic life of a new metal press. The press costs $10,000 initially. First-year maintenance costs are $1000. Maintenance costs are forecast to increase $1000 per year for each year after the first. Fill in the table and determine the economic life of the press. Consider only maintenance and capital recovery in your analysis. Interest is 5%.

Year	Maintenance Cost	EUAC of Capital Recovery	EUAC of Maintenance	Total EUAC
1	$1000			
2	2000			
...	...			
8	8000			

Solution

Year	Maintenance Cost	EUAC of Capital Recovery	EUAC of Maintenance	Total EUAC
1	$1000	$11,500	$1000	$12,500
2	2000	6,151	1465	7,616
3	3000	4,380	1907	6,287
4	4000	3,503	2326	5,829
5	5000	2,983	2723	5,706
6	6000	2,642	3097	5,739
7	7000	2,404	3450	5,854
8	8000	2,229	3781	6,010

Economic life = 5 years (EUAC = minimum)

EUAC of capital recovery = $10,000 (A/P, 15%, n)

EUAC of maintenance = $1000 + 1000 (A/G, 15%, n)

13-2

A robot's first cost is $40,000, and its market value declines by 20% annually. The operating and maintenance costs start at $2000 per year and climb by $1500 each year. If the MARR is 8%, find the minimum EUAC and the machine's economic life.

Solution

	A	B	C	D	E
1	8%	MARR			
2	$40,000	First Cost			
3	20%	Market value decline			
4	$2,000	O&M cost first year			
5	$1,500	Annual increase in O&M			
6					
7	Year	Cost	PWcost(t)	S(t)	EUAC
8	0	$40,000			
9	1	$2,000	$41,852	$32,000	$13,200
10	2	$3,500	$44,853	$25,600	$12,844
11	3	$5,000	$48,822	$20,480	$12,636
12	4	$6,500	$53,599	$16,384	**$12,547**
13	5	$8,000	$59,044	$13,107	$12,554
14	6	$9,500	$65,031	$10,486	$12,638
15	EUAC minimum =	**$12,547**			
16	Economic Life is	4 years			

13-3

A machine's first cost is $60,000 with salvage values over the next 5 years of $50K, $40K, $32K, $25K, and $12K. The annual operating and maintenance costs are the same every year. Determine the machine's economic life and its minimum EUAC, if the interest rate is 7%.

Solution

The O&M costs are constant so they can be ignored or included.

	A	B	C	D	E
1	7%	MARR			
2	$60,000	First Cost			
3	in table	Market value decline			
4	$0	O&M costs			
5	$0	Increase in O&M cost			
6					
7	Year	Cost	PWcost(t)	S(t)	EUAC
8	0	$60,000			
9	1	$0	$60,000	$50,000	$14,200
10	2	$0	$60,000	$40,000	$13,862
11	3	$0	$60,000	$32,000	$12,909
12	4	$0	$60,000	$25,000	**$12,083**
13	5	$0	$60,000	$12,000	$12,547
14	EUAC minimum =	**$12,083**			
15	Economic Life is	**4 years**			

13-4

A petroleum company, whose minimum attractive rate of return is 10%, needs to paint the vessels and pipes in its refinery periodically to prevent rust. Tuff-Coat, a durable paint, can be purchased for $8.05 a gallon, while Quick-Cover, a less durable paint, costs $3.25 a gallon. The labor cost of applying a gallon of paint is $6.00. Both paints are equally easy to apply and will cover the same area per gallon. Quick-Cover is expected to last 5 years. How long must Tuff-Coat promise to last to justify its use?

Solution

This replacement problem requires that we solve for a breakeven point. Let N represent the number of years Tuff-Coat must last. The easiest measure of worth to use in this situation is equivalent annual worth (EAW). Although more computationally cumbersome, other measures could be used and, if applied correctly, would result in the same answer.

Find N such that $EAW_{TC} = EAW_{QC}$

$$14.05(A/P, 10\%, N) = 9.25 \ (A/P, 10\%, 5)$$
$$(A/P, 10\%, N) = 0.17367$$

Searching the $i = 10\%$ table yields $N = 9$ years.

Tuff-Coat must last at least 9 years. Notice that this solution implicitly assumes that the pipes need to be painted indefinitely (i.e., forever) and that the paint and costs of painting never change (i.e., no inflation or technological improvements affecting the paint or the cost to produce and sell paint, or to apply the paint).

13-5

A hospital is considering the purchase of a new $40,000 diagnostic machine that will have no salvage value after installation, as the cost of removal equals any resale value. Maintenance is estimated to be $2000 per year as long as the machine is owned. After 10 years the machine must be scrapped because the radioactive ion source will have caused so much damage to machine components that safe operation is no longer possible. The most economic life of this machine is

a.   One year, since it will have no salvage after installation.

b.   Ten years, because maintenance doesn't increase.

c.   Less than 10 years, but more information is needed to determine the economic life.

Solution

The correct answer is b.

13-6

A graduate of an engineering economy course has compiled the following set of estimated costs and salvage values for a proposed machine with a first cost of $15,000; however, he has forgotten how to find the most economic life. Your task is to show him how to do this by calculating the equivalent annual cost (EUAC) for $n = 8$, given a MARR of 12%.

Life (n) Years	Estimated End-of-Year Maintenance	Estimated Salvage if Sold in Year n
1	$ 0	$10,000
2	$ 0	9,000
3	300	8,000
4	300	7,000
5	800	6,000
6	1300	5,000
7	1800	4,000
8	2300	3,000
9	2800	2,000
10	3300	1,000

Remember: Calculate only <u>one</u> EUAC (for $n = 8$). You are not expected to actually find the most economical life.

Solution

First cost   EUAC = 15,000(*A*/*P*, 12%, 8) = $3286.50

Salvage value   EUAC = −3000(*A*/*F*, 12%, 8) = −$243.90

Maintenance   EUAC = 300(*F*/*A*, 12%, 6)(*A*/*F*, 12%, 8)

$$+ 500(P/G, 12\%, 5)(P/F, 12\%, 3)(A/P, 12\%, 8) = \$1234.67$$

Total   EUAC$_8$ = $4010.27

(A complete analysis would show that the most economic life is 7 years, with EUAC = $3727.)

13-7

A truck salesperson is quoted as follows:

"Even though our list price has gone up to $42,000, I'll sell you a new truck for the old price of $40,000, an immediate savings of $2000, and give you a trade-in allowance of $21,000, so your cost is only ($40,000 − 21,000) = $19,000. The book value of your old truck is $12,000, so you're making an additional ($21,000 − 12,000) = $9000 on the deal." The salesperson adds, "Actually I am giving you more trade-in for your old truck than the current market value of $19,500, so you are saving an extra ($21,000 − 19,500) = $1500."

a.   In a proper replacement analysis, what is the first cost of your current truck?

b.   In a proper replacement analysis, what is the first cost of the new truck?

Solution

   a. $19,500

      The existing asset's first cost is always the current market value, not trade-in or book value.

   b. $40,000 is the offered price for the new truck.

      The inflated trade-in value of $1500 (21,000 − 19,500) may represent the salesperson's negotiable price, which would lower the price to $40,000 − $1500 = $38,500.

13-8

A car was purchased 4 years ago for $25,000. Its estimated salvage value after 7 years was $8000. The car can be sold for $14,000 now or $10,000, $7000, or $5000 in each later year. The annual maintenance cost will be $1800 for this year and increasing by $400 per year. What are the relevant cash flows for choosing how long to keep the car?

Solution

The relevant costs to consider are shown in the table.

Year	Market Value	O&M Costs
0	$14,000	
1	$10,000	$1800
2	$7,000	$2200
3	$5,000	$2600

13-9

Ten years ago Hyway Robbery installed a conveyor system for $8000. The conveyor system has been fully depreciated to a zero salvage value. The company is considering replacing the conveyor because maintenance costs have been increasing. The estimated end-of-year maintenance costs for the next 5 years are:

Year	Maintenance
1	$1000
2	1250
3	1500
4	1750
5	2000

At any time, the cost of removal just equals the value of the scrap metal recovered from the system. The replacement the company is considering has an EUAC of $1028 at its most economic life. The company has a minimum attractive rate of return (MARR) of 10%.

a. Should the conveyor be replaced now? Show the basis used for your decision.
b. If the old conveyor could be sold at any time as scrap metal for $500 more than the cost of removal and all other data remain the same, should the conveyor be replaced now?

Solution

a. Since the salvage value is not changing, each year's marginal cost equals the year's maintenance cost. In year 1 the existing conveyor's marginal cost is lower than the new asset's EUAC of $1028. In year 2 when the marginal cost increases to $1250, the conveyor should be replaced.

b.

Year	Cash Flow
0	−500
1	−1000
S	+500

$MC_1 = 1000 + 500(A/P, 10\%, 1) - 500 = \$1050$. Replace the conveyor now.

13-10

Ten years ago, the Cool Chemical Company installed a heat exchanger for $10,000. Maintenance costs have been increasing, and they will be $1000 this year. The cost of removal will be $1500 more than the heat exchanger is worth as scrap metal. The replacement the company is considering has an EUAC of $800 at its most economic life. If the company's minimum attractive rate of return (MARR) is 10%, should the heat exchanger be replaced now?

Solution

The marginal cost for this year for the existing heat exchanger is based on:

Year	Cash Flow	
0	+1500	Avoided salvage cost
1	−1000	Maintenance
S	−1500	Salvage cost

$MC_1 = -1500(A/P, 10\%, 1) + 1000 + 1500 = \$850$ → replace now since new exchanger's
    EUAC is $800

13-11

A firm uses a MARR of 12%. A crane was purchased 4 years ago for $180,000 and it has a current market value of $60,000. Expected operating and maintenance costs and market values follow. Data for a new crane have been analyzed. Its most economic life is 8 years with a minimum EUAC of $38,000. When should the existing crane be replaced?

Year	O&M cost	Market value
1	17,000	50,000
2	20,000	42,000
3	25,000	35,000
4	30,000	30,000
5	35,000	24,000

Solution

The marginal cost of the existing crane is higher in year 4 so it should be replaced after 3 years.

	A	B	C	D	E	F
1	12%	MARR				
2						
3	Year	Market Value	Loss in Market Value	Interest in Year n	O&M Cost	Marginal Costs
4	0	59,000				
5	1	50,000	9,000	7,080	17,000	33,080
6	2	42,000	8,000	6,000	20,000	34,000
7	3	35,000	7,000	5,040	25,000	37,040
8	4	30,000	5,000	4,200	30,000	39,200
9	5	24,000	6,000	3,600	35,000	44,600

13-12

One year ago, Machine *A* was purchased for $15,000, to be used for 5 years. The machine has not performed as expected, and it costs $750 per month for repairs, adjustments, and downtime. Machine *B*, designed to perform the same functions, can be purchased for $25,000 with monthly costs of $75. The expected life of machine *B* is 5 years. Operating costs are substantially equal for the two machines, and salvage values for both are negligible. If 6% is used, the incremental annual net equivalent of Machine *B* is nearest to

a. $2165
b. $2886
c. $4539
d. $5260

Solution

$$\Delta EUAW = -25{,}000(A/P, 6\%, 5) + 12(750 - 75)$$
$$= \$2165$$

The answer is a.

Chapter 14

Inflation and Price Change

14-1

A company requires a real MARR of 12%. What unadjusted MARR should be used if inflation is expected to be 2%?

Solution

Unadjusted MARR = $(1.12)(1.02) - 1 = 0.1424$ or 14.24%

14-2

The real interest rate is 4%. The inflation rate is 3%. What is the apparent interest rate?

Solution

$i = i' + f + i'$
$= 0.04 + 0.03 + 0.04(0.03) = 7.12\%$

14-3

A project that was analyzed under the assumption of 3% inflation was found to have an unadjusted internal rate of return (IRR) of 18%. What is the real IRR for the project?

Solution

Real IRR = $(1.18)/(1.03) - 1 = 0.1456$, or 14.56%

14-4

An electronic device cost $1250 in 2011. If inflation has averaged 2% each year, what is the price of the device in 2018?

Solution

$F = (1 + f)^n$
$F = (1 + 0.02)^7$
$= \$1435.86$

14-5

The apparent interest rate is 9.18% and the real interest rate is 6%. What is the inflation rate?

Solution

$$i = i' + f + i'f$$
$$0.0918 = 0.06 + f + 0.06f$$
$$0.0318 = 1.06f$$
$$f = 0.03$$

14-6

A government agency has predicted 3% inflation for the next 5 years. How much will an item that presently sells for $1000 cost in 5 years?

Solution

	A	B	C	D	E	F	G	H
1	i	n	PMT	PV	FV	Solve For	Answer	Formula
2	3.0%	5		$1,000		FV	$1,159	=-FV(A2,B2,C2,D2)

14-7

If prices have increased 50% over 10 years, what is the inflation rate?

Solution

	A	B	C	D	E	F	G	H
1	i	n	PMT	PV	FV	Solve For	Answer	Formula
2		8		-$1.00	$1.45	RATE	4.75%	=RATE(B2,C2,D2,E2)

14-8

An automobile that cost $24,500 in 2018 had an equivalent model 4 years later that cost $29,250. If the increase is attributed to inflation, what was the average annual rate of inflation?

Solution

$$F = P(1 + i_f)^n$$
$$29,250 = 24,500(1 + i_f)^4$$
$$\frac{29,250}{24,500} = (1 + i_f)^4$$
$$1 + i_f = (1.1939)^{1/4}$$
$$1 + i_f = 1.0453$$
$$i_f = 4.53\%$$

14-9

Jorge purchases a lot for $40,000 cash and plans to sell it after 5 years. What should he sell it for if he wants a 15% before-tax rate of return, after taking the 3% annual inflation rate into account?

Solution

$$F = 40,000(F/P, 15\%, 5)(F/P, 3\%, 5)$$
$$= \$93,268.56$$

14-10

The apparent interest rate is 9.18% and the real interest rate is 6%. What is the inflation rate?

Solution

$$i = i' + f + i'f$$
$$0.0918 = 0.06 + f + 0.06f$$
$$0.0318 = 1.06f$$
$$f = 0.03$$

14-11

The cost of a wastewater treatment plant for a small town of 6000 people was estimated to be about $85/person in 2006. If a modest estimate of the rate of inflation is 2.5% for the period to 2006, what will be the per capita cost of the treatment plant in 2021?

Solution

$$F = P(1 + i_f)^n$$
$$= 85(1 + 0.025)^{15}$$
$$= 85(1.4483)$$
$$= \$123.11$$

14-12

An investor is considering the purchase of a bond. The bond has a face value of $1000 and an interest rate of 6%; it pays interest once a year and matures in 8 years. This investor's real MARR is 25%. If the investor expects an inflation rate of 4% per year for the next 8 years, how much should he be willing to pay for the bond?

Solution

To earn a real 25% return with inflation of 4%, the nominal MARR must be equal to $(1.25)(1.04) - 1 = 30\%$,

$$NPV = 0 \text{ at IRR}$$
$$0 = -FC + 60(P/A, 30\%, 8) + 1000(P/F, 30\%, 8)$$
$$FC = \$250.50$$

14-13

A bond that pays no interest is called a zero-coupon bond. A $10,000 zero-coupon bond that matures in 10 years can be purchased today. If the expected annual rate of inflation is 3% and the buyer's unadjusted MARR is 8%, what is the maximum that should be paid for the bond?

Solution

$i = i' + f + i'f$

$= 0.08 + 0.03 + 0.08(0.03)$

$= 0.1124$

$P = 10,000(1 + 0.1124)^{-10}$

$= \$3446.59$

14-14

Sylvia B. bought an 8% tax-free municipal bond. The cost of the bond was $1000, and it will pay $80 each year for 20 years. The bond will mature at that time, returning the original $1000. If inflation is expected to average 3% during the period, what is the inflation-adjusted rate of return?

Solution

$i = 8\%$ $f = 3\%$

$i = i' + f + i'f$

$0.08 = i' + 0.03 + i'(0.03)$

$i' = 0.0485 = 4.85\%$

14-15

An inventor expects to receive $75,000 per year from royalties on a patent for 12 years. If the inventor's interest rate is 9% and the inflation rate is 4%, determine the present value of the cash flows.

Solution

The PV is $672,013

	A	B	C	D	E	F	G
1	inflation	4.0%					
2	interest	9.0%					
3	i'	4.808%					
4	i' = (i - f)/(1 + f)						
5	i	n	PMT	PV	FV	Solve For	Answer
6	4.81%	12	$75,000			PV	-$672,013

14-16

A vacant lot is purchased for $20,000. After 5 years the lot is to be offered for sale. If the buyer requires a before-tax return on investments of 15% and inflation has averaged 4% per year over the 5-year period, what is the required selling price?

Solution

$F = 20,000(F/P, 4\%, 5)(F/P, 15\%, 5)$

$= \$48,947.74$

14-17

A product has sales of $7M this year, but sales are expected to decline at 10% per year until it is discontinued after year 5. If the firm's interest rate is 15%, calculate the PW of the revenues.

Solution

⊿	A	B	C
1	15%	interest rate	
2	-10.0%	Rev. change	
3			
4	Year	Revenue	
5	1	$7,000,000	
6	2	$6,300,000	
7	3	$5,670,000	
8	4	$5,103,000	
9	5	$4,592,700	
10	6	$4,133,430	
11	PW	$21,566,815	=NPV(A1,B5:B10)
12	Note there is no time 0 cash flow.		

14-18

How much life insurance should a person buy if he wants to leave enough money to ensure that his family through generations will receive $25,000 per year in interest, of constant Year-0-value dollars? The interest rate expected from banks is 11%, while the inflation rate is expected to be 4% per year.

Solution

The actual (effective) rate that the family will be getting is

$$i' = \frac{i-f}{1+f} = \frac{0.11-0.04}{1.04} = 0.0673 = 6.73\%$$

To calculate P, $n = \infty$ (capitalized cost)

$$P = \frac{A}{i'} = \frac{25,000}{0.0673} = \$371,471$$

Therefore, he needs to buy about $371,500 of life insurance.

14-19

An electronics store offers two options to buy a new laptop computer that has a price of $440. A customer can pay cash and immediately receive a discount of $49, or she can pay for the computer on an installment plan. The installment plan has a nominal rate of 12% compounded bi-yearly and requires an initial down payment of $44, followed by four equal payments (principal and interest) every 6 months for 2 years. If for the typical customer the cost of money is 5%, what is the maximum effective annual inflation rate for the next 2 years that would make paying cash preferable to paying installments? All figures given are quoted in Time-0 dollars.

Solution

If the installment plan were selected, the monthly payments in nominal dollars would be

$$(-440 + 44)(A/P, 6\%, 4) = -\$114.28$$

The breakeven inflation rate is that such that

$$NPV_{BUY} = NPV_{INSTALL} \text{ or } NPV_{BUY\ -\ INSTALL} = 0$$

$$NPV_{B-I} = [(-440 + 49) + 44] + 114.28(P/A, i_{1/2}, 4) = 0$$

Since $(P/A, i_{1/2}, 4) = 3.0364$, the nominal effective semiannual cost of money would have to be $i_{1/2} = 0.115$. The nominal effective annual rate would be $i = (1.115)^2 - 1 = 0.2432$.

The effective annual inflation rate can now be computed from the formula

$$(1.2432) = (1.05)\ (1 + f)$$
$$f = 0.1840$$

14-20

A machine has a first cost of \$100,000 (in today's dollars) and a salvage value of \$20,000 (in then-current dollars) at the end of its 10-year life. Each year the machine is used will eliminate the job of one full-time worker. A worker costs \$30,000 (today's dollars) in salary and benefits. Labor costs are expected to escalate at 10% per year. Operating and maintenance costs will be \$10,000 per year (today's dollars) and will escalate at 7% per year.

Construct a table showing before-tax cash flows in current dollars and in today's dollars. The inflation rate is 7%.

Solution

End of	Current Dollars				Today's
Year	Savings	O&M	Capital	Total	Dollars
0			−100,000	−100,000	−100,000
1	33,000	−10,700		22,300	20,841
2	36,300	−11,449		24,851	21,706
3	39,930	−12,250		27,680	22,595
4	43,923	−13,108		30,815	23,509
5	48,315	−14,026		34,290	24,448
6	53,147	−15,007		38,140	25,414
7	58,462	−16,058		42,404	26,407
8	64,308	−17,182		47,126	27,428
9	70,738	−18,385		52,354	28,477
10	77,812	−19,672	20,000	78,141	39,723

14-21

On January 1, 2005, the National Price Index was 208.5, and on January 1, 2015, it was 516.71. What was the inflation rate, compounded annually, over that 10-year period? Assuming that the same rate continued to hold for the next 10 years, what would the National Price Index be on January 1, 2025?

Solution

Set NPW = 0

$$0 = -208.5 + 516.71(P/F, i_f, 10)$$

$$(P/F, i_f, 10) = \frac{208.5}{516.71}$$

$$= 0.4035$$

From interest tables, the P/F factor at   9% = 0.4224

$$10\% = 0.3855 \qquad \therefore \ 9\% < i_f < 10\%$$

By interpolation, $i_f = 9.51\%$.

National Price Index$_{2025}$ = $516.71(1 + 0.0951)^{10}$ = 1281.69

14-22

A firm's internal price index for the cost of the raw materials for its most important product was 175 four years ago. Those prices have averaged an annual increase of 7% since. Calculate the current value of the index.

Solution

	A	B	C	D	E	F	G	H
1	i	n	PMT	PV	FV	Solve For	Answer	Formula
2	7.0%	4		175		FV	229	=-FV(A2,B2,C2,D2)

14-23

Your savings account pays 4% interest on the $40,000 you deposited at time 0. Inflation was 3% for 4 years and then 2% for 6 years. How much is in the account after 10 years in year-10 dollars? How much is that in year-0 dollars? What has been your real rate of return?

Solution

	A	B	C	D	E	F	G	H
1	i	n	PV	PMT	FV	Solve For	Answer	
2	4.0%	10	-$40,000			FV	$59,210	in yr-10 $s
3	3.0%	4			-$59,210	PV	$52,607	
4	2.0%	6			-$52,607	PV	$46,714	in yr-0 $s
5		10	-$40,000		$46,714	RATE	1.56%	real return

14-24

A European investor lives near one of his country's borders. In Country A (where he lives), banks are offering an 8% interest rate and the inflation rate is 3%. Country B, on the other hand, has an inflation rate of 23%, and banks are offering 26% interest on deposits.

a. What real or effective interest rate does the investor earn when investing in his own country?
b. The investor believes that the currency of Country B will not change in its value relative to the value of the currency of Country A during this year. In which country would he get a larger effective interest rate?
c. Suppose that he invests in a bank in Country B and that his prediction turns out to be wrong: the currency of Country B was devaluated 20% with respect to the exchange value of Country A's currency. What effective interest rate would he obtain in this case?

Solution

a. $i' = ?$ if $i = 80\%$, and $f = 3\%$
$$i = i' + f + i'f$$
$$0.08 = i' + 0.03 + i'(0.03)$$
$$i' = 0.0485$$
$$= 4.85\%$$

b. If investment in Country A: $i'_A = 0.0485$
If investment in Country B: $i_B = 26\%, f_A = 3\%$
(The investor <u>lives</u> in Country A; inflation of Country B has no direct affect on him.)
$$i'_B = \frac{i_B - f_A}{1 + f_A} = \frac{0.26 - 0.03}{1 + 0.03} = 0.2233 = 22.33\%$$

He can get a larger effective interest rate in Country B.

c. Let $X =$ amount originally invested in B (measured in currency A).
The amount collected at end of 1 year (measured in currency A) is
$$\underbrace{(1.0 - 0.2)}_{\substack{\text{Due to the} \\ \text{devaluation}}} \times \underbrace{(1.26)}_{\substack{\text{Due to initial} \\ \text{deposit (+)interest}}} = 1.008X$$

The interest is then $i = \dfrac{1.008X - X}{X} = 0.008$.

During the year inflation in Country A (where the investor lives) was 3%; therefore,
$$i = 0.008$$
$$f = 0.03$$
$$i' = ?$$
$$i' = \frac{0.008 - 0.03}{1 + 0.03} = -0.02136$$

He actually would have lost money (negative effective interest rate of -2.136%).

14-25

Property, in the form of unimproved land, is purchased at a cost of $8000 and is held for 6 years, at which time it is sold for $32,600. During each of the six years of ownership $220 is paid in property tax and may be accounted for at an interest of 12%. The income tax rate on the long-term capital gain is 15% of the gain. Inflation during the period is 4% per year. What is the annual rate of return for this investment?

Solution

Long-term gains = 32,600 − 8000 = 24,600
Tax on long-term gain = 0.15 × 24,600 = 3690
Property tax = 220(F/A, 12%, 6) = $1785.30
Adjusted FW = 32,600 − 3,690 − 1785.30 = $27,624.70
also FW = $8000(1 + i_{eq})^6$

$$\therefore \left(1 + i_{eq}\right) = \left(\tfrac{27,624.70}{8000}\right)^{\frac{1}{6}} = 1.2294$$

$$\left(1 + i_{eq}\right) = \left(1 + i\right)\left(1 + i_f\right)$$

$$1 + i = \frac{1.2294}{1.04} = 1.1821 \quad \text{or } 18.2\% \quad \text{rate of return}$$

14-26

A lot purchased for $4500 is held for 5 years and sold for $13,500. The average annual property tax is $45 and may be accounted for at an interest rate of 12%. The income tax rate on the long-termcapital gain is 15% of the gain. What is the rate of return on the investment if the allowance for inflation is treated at an average annual rate of 6%?

Solution

Long-term gain = 13,500 − 4,500 = 9,000
Tax on long-term gain = (0.15)(9000) = 1350
Property tax = 45(F/A, 12%, 5) = 285.89
Adjusted FW = 13,500 − 1350 − 285.89 = 11,864.12
also FW = $4500(1 + i_{eq})^5$

$$\therefore \left(1 + i_{eq}\right) = \left(\tfrac{11,864.12}{4500}\right)^{\frac{1}{5}} = 1.214$$

$$\left(1 + i_{eq}\right) = \left(1 + i\right)\left(1 + i_f\right)$$

$$1 + i = \frac{1.214}{1.06} = 1.1453 \quad \text{or } 14.5\% \text{ rate of return}$$

14-27

Undeveloped property near the planned site of an interstate highway is estimated to be worth $48,000 in 6 years when the construction of the highway will be completed. Consider a 15% capital gains tax on the gain, an annual property tax of 0.85% of the purchase price, an annual inflation rate of 7%, and an expected return of 15% on the investment. What is the indicated maximum purchase price now?

Solution

Let X = purchase cost

$1 + i_{eq} = (1.15)(1.07) = 1.231$

Annual property tax = $0.0085X$

FW of property tax $= 0.0085X(F/A, 23.1\%, 6) = 0.0909X$

Adjusted return = $48,000 - 0.15(48,000 - X) - 0.0909X$

Also $= X(1.231)^6 = 3.48X$

Therefore, $40,800 + 0.15X - 0.0909X = 3.48X$

$$X = \$11,927 \text{ purchase price}$$

14-28

A solar heating system costs $6500 initially and qualifies for a federal tax credit (40% of cost, not to exceed $4000). The cost of money is 10%, and inflation is expected to be 7% during the life of the system. The expected life of the system is 15 years with zero salvage value. The homeowner is in the 40% income tax bracket. The initial (first-year) fuel saving, estimated at $500, is expected to increase in response to inflation. The annual maintenance cost of the system is established at 5% of the annualized cost of the system. What is the time required for the payback condition to be reached for this investment?

Solution

Adjust initial cost by tax credit: $P = 0.60(6500) = 3900$

Annualized cost: $A = 3900(A/P, 10\%, 15) = 512.85$

$1 + i_c = 1.10[1 + 40(0.10)]/1.05 = 1.0895$

$1 + i_m = 1.05$ represents maintenance charge as a rate

PW of costs $= 512.85(P/A, 8.95\%, 15) = 512.85(8.086) = 4146.67$

$1 + i_{eq} = (1 + i)/(1 + i_f) = 1.10/1.07 = 1.028$

Try 9 years: PW = $500\ (P/A, 2.8\%, n) = 500(7.868) = \$3934.18$

Try 10 years: PW = $500\ (P/A, 2.8\%, n) = 500(8.618) = \$4308.97$

9 years < Payback < 10 years   By interpolation, Payback = 9.6 years

14-29

The net cost of a home solar heating system, expected to last for 20 years, is $8000. If the value of money is 10%, inflation is expected to be 8%, and the initial annual fuel saving is $750, what is the time for the payback condition to be reached for the system? Assume that the homeowner is in the 30% income tax bracket.

Solution

Annualize P: $A = 8000 \, (A/P, 10\%, 20) = 940$

$1 + i_c = (1.10)[1 + 0.10(0.30)] = 1.133$

PW of cost $= 940(P/A, 13.3\%, 20) = 940(6.900) = 6486$

$1 + i_{eq} = (1 + i)/(1 + i_f) = 1.10/1.08 = 1.0185$

Try 9 years: PW $= 940(P/A, 1.85\%, n) = \$6171$

Try 10 years: PW $= 940(P/A, 1.85\%, n) = \$6790$

9 years < Payback < 10 years   By interpolation, Payback = 9.5 years.

Chapter 15

Selection of a Minimum Attractive Rate of Return

15-1

The capital structure of a firm is as follows.

Source of Capital	Share of Capitalization	Interest Rate
Loans	35	7%
Bonds	40	8%
Common stock	25	10%

The combined state and federal income tax rate for the firm is 42%. Find the after-tax and before-tax costs of capital to the firm.

Solution

Before-tax cost of capital

$(0.35 \times 7\%) + (0.40 \times 8\%) + (0.25 \times 10\%) = 8.15\%$

After-tax cost of capital

$(0.35 \times 7\%)(1 - 0.42) + (0.40 \times 8\%)(1 - 0.42) + (0.25 \times 10\%) = 5.78\%$

15-2

NuStuff Inc. has raised $8M by selling bonds at an average rate of 7%. NuStuff's stockholders expect a 14% rate of return, and there is $15M in common stock and retained earnings. NuStuff has $4M in loans at an average rate of 9%. What is the firm's cost of capital:

(*a*) Before taxes?

(*b*) After taxes with a combined tax rate of 26%?

Solution

	A	B	C	D	E	F
1	26%	tax				
2	Source	Amount	Rate	col. B x C	AT Rate	col. B x E
3	Bond	$8	7%	$0.56	5.18%	$0.41
4	Loan	$4	9%	$0.36	6.66%	$0.27
5	Stock	$15	14%	$2.10	14.00%	$2.10
6		$27		$3.02		$2.78
7						
8		Cost of Capital		11.19%		10.30%

15-3

Abby Industries Inc., has the following capital structure:

Type	Amount	Average Minimum Return
Mortgages	$ 25,000,000	7%
Bonds	180,000,000	9%
Common stock	100,000,000	10%
Preferred stock	50,000,000	8%
Retained earnings	120,000,000	10%

Determine the weighted average cost of capital (WACC) for Abby.

Solution

$$WACC = \frac{25M(.07) + 180M(.09) + 100M(.10) + 50M(.08) + 120M(.10)}{(25M + 180M + 100M + 50M + 120M)} = 9.25\%$$

15-4

A new utility had $10M in start-up capital, and it has sold $45M in bonds at 7% to raise money for electric power generation and connecting to the existing power grid. There were some cash flow problems so it had to borrow $2.5M at 11%. The stockholders expect a 15% rate of return, and the combined tax rate is 25%. What is the utility's before- and after-tax costs of capital?

Solution

	A	B	C	D	E	F
		25% tax		Before Tax		After Tax
	Source	Amount	Rate	col. B x C	AT Rate	col. B x E
	Bond	$45.0	7%	$3.15	5.25%	$2.36
	Loan	$2.5	11%	$0.28	8.25%	$0.21
	Stock	$10.0	15%	$1.50	15.00%	$1.50
		$58		$4.93		$4.07
		Cost of Capital		8.57%		7.08%

15-5

A loan has an interest rate of 8%, and the inflation rate is 2.4%. What is the loan's real interest rate adjusted for inflation?

Solution

Applying Eq. 14-1:

Loan rate = Inflation rate + Real rate + Inflation rate × Real rate

Rearrange to arrive at:

Real rate = (Loan rate − Inflation rate)/(1 + Inflation rate) = (.08 − .024)/1.024 = .055 = 5.5%

15-6

A country with an inflation rate of 20% is offering 5-year bonds at an interest rate of 30%. What is the real rate on the bond? Appendix 9A and Figure 13-1 use a common approximation for the real rate. What is that value in this case?

Solution

Applying Eq. 14-1:

Bond rate = Inflation rate + Real rate + Inflation rate × Real rate

Rearrange to arrive at:

Real rate = (Bond rate − Inflation rate)/(1 + Inflation rate) = (.30 − .20)/1.20 = .083 = 8.3%

Approximation is:

Real rate ≈ Bond rate − Inflation rate = .3 − .2 = 10%

15-7

A small surveying company identifies its available independent alternatives.

	Alternative	Initial Cost	Rate of Return
A:	Repair existing equipment	$1000	30%
B:	Buy EDM instrument	2500	9%
C:	Buy a new printer	3000	11%
D:	Buy equipment for an additional crew	3000	15%
E:	Buy programmable calculator	500	25%

The owner of the company has $5000 of savings currently invested in securities yielding 8% that could be used for the company.

a. Assuming that the funds are limited to the owner's savings, what is the apparent cutoff rate of return?

b. If the owner can borrow money at 10%, how big a loan should she undertake?

Solution

a.

Alt.	Investment	Cumulative Investment	IRR	
A	$1000	$ 1,000	30%	
E	500	1,500	25%	
D	3000	4,500	15%	← Cutoff rate of return
C	3000	7,500	11%	= 11–15%
B	2500	10,000	9%	

b. Do all projects with a rate of return exceeding 10%. Thus Alternatives A, E, D, and C, with a total initial cost of $7500, would be selected. Since only $5000 is available, $2500 would need to be borrowed.

15-8

Barber Brewing is in the process of determining the capital budget for the coming year. The following projects are under consideration.

	A	B	C	D
First cost	$10,000	$13,000	$20,000	$33,000
Annual income	10,000	9,078	16,000	16,455
Annual cost	7,362	5,200	11,252	7,300

All projects have a 5-year useful life. If Barber's budget is set at $50,000, which alternative(s) should be selected?

Solution

NPW = 0 at IRR.

$0 = -FC + $ Net Income $(P/A, i, 5)$ Therefore $(P/A, i, 5) = FC/$Net Income

			Rank
IRR_A	$P/A = 10,000/2638 = 3.791$	IRR = 10%	3
IRR_B	$P/A = 13,000/3878 = 3.352$	IRR = 15%	1
IRR_C	$P/A = 20,000/4748 = 4.212$	IRR = 6%	4
IRR_D	$P/A = 33,000/9155 = 3.605$	IRR = 12%	2

Choose projects B and D. Total = $13,000 + 33,000 = $46,000.

15-9

A manufacturing plant has a capital budget of $350,000. Which of these projects should be done?

Project	First Cost	Annual Benefit	Life (years)
1	$200,000	$50,000	15
2	$50,000	$7,000	20
3	$100,000	$40,000	6
4	$50,000	$15,000	7
5	$150,000	$30,000	15

Solution

	A	B	C	D	E	F
1	Project	First Cost	Annual Benefit	Life (years)	IRR	Cumulative First Cost
2	3	$100,000	$40,000	6	32.66%	$100,000
3	1	$200,000	$50,000	15	24.01%	$300,000
4	4	$50,000	$15,000	7	22.93%	$350,000
5	5	$150,000	$30,000	15	18.42%	$500,000
6	2	$50,000	$7,000	20	12.72%	$550,000

Do Projects 3, 1, & 4.

15-10

Which projects should be done, if a parks and recreation department has a capital budget of $350,000?

Project	First Cost	Annual Benefit	Life (years)
1	$50,000	$12,000	15
2	$50,000	$9,000	20
3	$100,000	$15,000	20
4	$50,000	$12,000	10
5	$100,000	$30,000	10
6	$100,000	$20,000	15
7	$150,000	$25,000	15

Solution

	A	B	C	D	E	F
			Annual	Life		Cumulative
1	Project	First Cost	Benefit	(years)	IRR	First Cost
2	5	$100,000	$30,000	10	27.32%	$100,000
3	1	$50,000	$12,000	15	22.91%	$150,000
4	4	$50,000	$12,000	10	20.18%	$200,000
5	6	$100,000	$20,000	15	18.42%	$300,000
6	2	$50,000	$9,000	20	17.25%	$350,000
7	7	$150,000	$25,000	15	14.47%	$500,000
8	3	$100,000	$15,000	20	13.89%	$600,000

Do Projects 5, 1, 4, &6.

15-11

A small construction company identifies the following alternatives, which are independent except where noted.

Alternative	Initial Cost	Incremental Rate of Return	On Investment Over
1. Repair bulldozer	$5,000	30.0%	0
2. Replace backhoe			
with Model A	20,000	15.0%	0
with Model B	25,000	10.5%	2A
3. Buy new dump truck			
Model X	20,000	20.0%	0
Model Y	30,000	14.0%	3X
4. Buy computer			
Model K	5,000	12.0%	0
Model L	10,000	9.5%	4K

a. Assuming that the company has $55,000 available for investment and is not able to borrow money, what alternatives should be chosen, and what is opportunity coast of capital?

b. If, however, the company can borrow money at 10%, how much should be borrowed, and which alternatives should be selected?

Solution

Rank the alternatives by ΔROR:

Project	Incremental Investment	Cumulative Investment	ΔIRR
1	$ 5,000	$ 5,000	30.0%
3X	20,000	25,000	20.0%
2A	20,000	45,000	15.0%
3Y – 3X	10,000	55,000	14.0%
4K	5,000	60,000	12.0%
2B – 2A	5,000	65,000	10.5%
4L – 4K	5,000	70,000	9.5%

a. With $55,000 available, choose projects:

1	Repair bulldozer
2A	Backhoe Model A
3Y	Dump truck Model Y
	No computer

Opportunity cost of capital = 14%.

b. Borrow $10,000. Choose projects:

1	Repair bulldozer
2B	Backhoe Model B
3Y	Dump truck Model Y
4K	Computer Model K

15-12

A variant of a firm's leading product was developed for a related market segment dominated by a competitor. The competitor's new product release is increasing its dominance in that segment. The revised plan is to release the product in a foreign market that the firm wants to enter. Should the firm go ahead or develop another plan? $45M has been spent on product development. The product will cost $50M to adapt and to introduce. The net sales should be $15M per year for 5 years. Good will from the product will be worth $20M in 5 years for new products in this market.

Solution

From Table 15-2 the required risk adjusted MARR is 20%. The $45M that has been spent is a sunk cost that should be ignored. The IRR of 22.3% exceeds the 20% required for a new product in a foreign market. Proceed.

	A	B	C	D	E	F	G
1	i	N	PMT	PV	FV	Solve For	Answer
2		5	$15	-$50	$20	RATE	22.3%

15-13

A new product is being developed for the normal market for this product line. Using Table 15-2 h.ow much higher do the annual net sales have to be in a related market to justify releasing the product there? In both cases another $5.4M is required, in neither case is there a salvage value, and in both cases sales are expected for 5 years.

Solution

	A	B	C	D	E	F	G	H
1	Market	i	N	PMT	PV	FV	Solve For	Answer
2	normal	10%	5		-$5.4	$0	PMT	$1.42
3	related	12%	5		-$5.4	$0	PMT	$1.50

From Table 15-2 the required risk adjusted MARRs are 10% & 12%. Annual sales must be 0.08M higher in the related market.

15-14

A city engineer calculated the present worth of benefits and costs of a number of possible projects, based on 10% interest and a 10-year analysis period.

Costs and Benefits ($1000s)

Project:	A	B	C	D	E	F	G
Present Worth of Costs	75	70	50	35	60	25	70
Present Worth of Benefits	105	95	63	55	76	32	100

If 10% is a satisfactory minimum attractive rate of return (MARR), and $180,000 is available for expenditure, which project(s) should be selected? Use a NPW/C ranking.

Solution

Project:	A	B	C	D	E	F	G
Present Worth of Costs	75	70	50	35	60	25	70
Present Worth of Benefits	105	95	63	55	76	32	100
NPW	30	25	13	20	16	7	30
NPW/C	.400	.357	.260	.571	.267	.280	.428
Rank	3	4	7	1	6	5	2

$D + G + A = 35K + 70K + 75K = \$180K$

Choose Projects D, G, and A.

15-15

The CFO of Republic Express has asked regional managers to submit capital projects for next year. The CEO has decided to fund each region's top request and to fund two additional requests. But no region may have more than two projects funded. Which projects should be funded and what is capital budget?

Region	Project	Cost	Annual Benefit	Life (years)
Southeastern (SE)	A	$ 90,000	$16,400	15
	B	40,000	15,000	5
	C	60,000	20,400	5
	D	120,000	27,600	20
Midwest (MW)	A	50,000	10,000	20
	B	120,000	36,700	15
	C	75,000	21,600	5
	D	50,000	16,200	5
Northeastern (NE)	A	50,000	16,700	20
	B	80,000	23,500	5
	C	75,000	26,100	10
Western (W)	A	60,000	16,900	15
	B	50,000	15,300	10

Solution

Region	Project	Cost	IRR
Southeastern (SE)	A	$ 90,000	16.3%
	B	40,000	**25.4**
	C	60,000	20.8
	D	120,000	22.6
Midwest (MW)	A	50,000	19.4
	B	120,000	**30.0**
	C	75,000	13.5
	D	50,000	18.6
Northeastern (NE)	A	50,000	**33.3**
	B	80,000	14.4
	C	75,000	**32.8**
Western (W)	A	60,000	**27.4**
	B	50,000	**28.0**

Highest in each region: SE = B, MW = B, NE = A, W = B. The highest IRRs unchosen NE = C and W = A.

Capital budget = 40,000 + 120,000 + 50,000 + 50,000 + 75,000 + 60,000 = $395,000

Chapter 16

Economic Analysis in the Public Sector

16-1

A city's park department has a capital budget of $600,000. What is the government's opportunity cost of capital and which projects should be done?

Project	First Cost	Return	B/C
A	$300,000	23%	1.45
B	$100,000	22%	1.50
C	$100,000	17%	1.40
D	$200,000	19%	1.23
E	$100,000	18%	1.26

Solution

Ranking on IRR shows that a budget of $600,000 has an opportunity cost of capital of 19%. Projects A, B, and D should be done.

	A	B	C	D	E
1	Project	First Cost	Return	B/C	Cumulative First Cost
2	A	$300,000	23%	1.45	$300,000
3	B	$100,000	22%	1.50	$400,000
4	D	$200,000	19%	1.23	$600,000
5	E	$100,000	18%	1.26	$700,000
6	C	$100,000	17%	1.40	$800,000

16-2

Tumbleweed Junction bought new traffic enforcement equipment for $18,000. It will annually generate revenues of $25,000 and expenses of $15,000 for 5 years. There is no salvage value. Compute the benefit/cost ratio if $i = 10\%$.

Solution

$PW_{BENEFITS}$ $= 25,000(P/A, 10\%, 5) = \$94,775$
PW_{COSTS} $= 15,000(P/A, 10\%, 5) + 18,000 = \$74,865$
B/C $= 94,775/74,865$
    $= 1.2659$

16-3

Froggy University may buy a new garbage incinerator. The "best" alternative costs $55,000 and is expected to save $6000 in garbage fees during the first year, with the savings increasing by $750 each year thereafter. The incinerator will decrease air quality around campus, which is estimated to be worth $1000 per year. The incinerator will have no salvage value at the end of its 10-year useful life. If Froggy U. evaluates all capital outlays with a 6% interest and requires B/C analysis, what would you recommend?

Solution

Consider the air quality value to be a disbenefit.

$$B/C = \frac{6000 + 750(A/G, 6\%, 10) - 1000}{55,000(A/P, 6\%, 10)} = 1.07 \qquad (>1) \qquad \therefore \text{ Purchase the incinerator.}$$

16-4

A new electric generation plant is expected to cost $43,250,000 to complete. Expected revenues will be $3,875,000 per year, while operational expenses will be $2,000,000 per year. If the plant is expected to last 40 years and the electric authority uses 3% as its cost of capital, determine whether the plant should be built.

Solution

$AW_B = \$3,875,000$
$AW_C = 43,500,000(A/P, 3\%, 40) + 2,000,000 = \$3,872,725$

$$B/C = \frac{AW_B}{AW_C} = \frac{3,875,000}{3,872,725} = 1 \qquad \therefore \text{ Build the plant.}$$

16-5

At 10% interest, what is the benefit/cost ratio for the following government project?

Initial cost	$2,000,000
Additional costs at end of Years 1 & 2	30,000
Benefits at end of Years 1 & 2	0
Annual benefits at end of Years 3–10	90,000

Solution

PW_{COST} $= 200K + 30K(P/A, 10\%, 2) = 252,080$

$PW_{BENEFITS}$ $= 90K(P/A, 10\%, 8)(P/F, 10\%, 2) = 396,800$

B/C $= 396,800/252,080$

$= 1.574$

16-6

Gordon City is considering a new garbage dump on the outskirts of town. The land will cost $85,000. Earthwork and other site preparation will cost $250,000. Environmental inspection prior to use will cost $15,000. The upkeep and operating costs will be $50,000 per year for its 8-year life. The new dump will reduce annual disposal fees by $6 each for the 27,000 customers it will serve. Assume that the number of customers will remain relatively constant during the dump's life. Changes in the environmental conditions adjacent to the dump will result in an annual cost of $32,000. At the end of its useful life the dump must be "capped" at a cost of $75,000. Determine the benefit/cost ratio if Gordon City uses 4% as its cost of money.

Solution

$AW_C = 350,000(A/P, 4\%, 8) + 50,000 + 75,000(A/F, 4\% 8)$

$= \$130,000$

$AW_B = (6 \times 27,000) - 32,000$ 32,000 is considered a disbenefit (reduction in benefits)

$= \$110,112.50$

B/C ratio $= 130,000/110,112.50$

$= 1.18$

16-7

A new water treatment plant will cost Frogjump $2,000,000 dollars to build and $100,000 per year to operate. After 20 years the salvage value will be 0. The new plant is more efficient; the utility bills for each of Frogjump's 6000 customers will drop $50 per year. The annual cost of reduced air quality is $5 per resident per year. The population of Frogjump, which is currently 18,000, is expected to remain relatively constant over the life of the plant. If 3% is used to evaluate public works projects, should the water treatment plant be built?

Solution

First cost: $2,000,000

Annual operating cost: $100,000/year

Note that # Residents > # customers

Air quality annual cost: $5/year × 18,000 residents = $90,000/year (considered a disbenefit)

Annual benefits: $50/year × 6000 customers = $300,000/year

$AW_B = \$300,000 - 90,000 = 210,000$

$AW_C = \$2,000,000(A/P, 3\%, 20) + 100,000 = \$234,400$

  $B/C = 210,000/234,400 = 0.8958$

$B/C < 1$; therefore, the water treatment plant should not be built.

16-8

Drygulch may dam the nearby Twisted River to create a recreational lake. The earthen dam will cost $1,000,000. Every 10 years dam reworking will cost $100,000. Annual operating expenses will be $20,000. Expected annual usage is 8000 persons with a monetary benefit to each user of $7.75. A fee of $6 will be charged to each of the boats that launch into the lake; an annual total of 1200 boats is estimated. At an interest rate of 4%, determine whether Drygulch should build the dam.

Solution

$AW_B = (8000 \times 7.75) - (1200 \times 6.00)$

  $= \$54,800$

$AW_C = 1,000,000(A/P, 4\%, \infty) + 100,000(A/F, 4\%, 10) + 20,000$

  $= \$68,330$

$B/C = 54,800/68,330$

  $= 0.80$ ∴ Do not build the dam.

16-9

The Tennessee Department of Highways (TDOH) is considering its first "tolled bypass" around Greenfield. The initial cost of the bypass will be $5.7 million. Installing tollbooths will cost $1.3 million. Annual bypass maintenance will be $105,000 while annual tollbooth costs will be $65,000. The bypass should increase tax revenues by $225,000 per year. TDOH projects user benefits to be $100,000 more than the tolls each year. Each user will pay a toll of $0.90. TDOH estimates that there will be 500,000 users annually. Other relevant data are as follows:

Resurfacing cost (every 7 years)   4% of bypass initial cost

Shoulder grading/rework   90% of resurfacing cost

If the state uses an interest rate of 7%, should the "tolled bypass" be constructed? Assume perpetual life.

Solution

Use benefit/cost analysis. (Use AW analysis.)

Benefits

Annual tax revenues		$225,000
Bypass user annual benefits		$100,000
Tollbooth revenue	$0.90(500,000)	$450,000
		$AW_B = $ 775,000

Costs

Tollbooth first cost	$1,300,000(A/P, 7\%, \infty)$	$ 91,000
Bypass first cost	$5,700,000(A/P, 7\%, \infty)$	$399,000
Tollbooth maintenance		$ 65,000
Bypass maintenance		$105,000
Resurfacing $0.04($5,700,000(A/P, 7\%, 7))$		$ 26,537
Shoulder grading/rework $0.90\{0.04[$5,700,000(A/P, 7\%, 7)]\}$		$ 23,721
		$AW_C = $710,258

B/C = 775,000/710,258 = 1.09

∴ The bypass should be constructed.

16-10

Mathis City may construct a municipal park. Land required would be bought as an initial $62,000 purchase and a $24,000 expansion 5 years later. Initial construction would take 2 years and cost $250,000 each year. The expansion cost at the start of year 6 would be $80,000. Park activities, such as putt-putt golf, would generate user fees of $35,000 per year. The monetary value of citizens' enjoyment is $26,000 per year. Maintenance and upkeep would be contracted out at a cost of $12,000 per year. The park would be used indefinitely, and Mathis City uses an interest rate of 8%. Determine the park's B/C ratio.

Solution

First calculate the PW of costs before converting to AW at $n = \infty$.

$PW_C = 62,000 + 24,000(P/F, 8\%, 5) + 250,000(P/F, 8\%, 1) + 250,000(P/F, 8\%, 2)$
$\qquad + 80,000(P/F, 8\%, 6)$
$\qquad = \$574,550$

$AW_C = 574,550(A/P, 8\%, \infty) + 12,000$
$\qquad = \$57,964$

$AW_B = 35,000 + 26,000$
$\qquad = \$61,000$

$\quad$ B/C = 61,000/57,964 = 1.0524

16-11

Two different water delivery methods are available to supply the town of Dry-Hole with much-needed water. Use an interest rate of 6% and the appropriate analysis for public projects to determine which should be chosen.

	Deep Well	Canal
First cost	$435,000	$345,000
Annual M&O costs	18,000	25,500
Useful life	20 years	20 years

Solution

$$B/C_{DW-C} = \frac{7500(P/A, 6\%, 20)}{90,000} = 0.96 \qquad (<1) \qquad \therefore \text{ Build the canal}$$

16-12

A city engineer is evaluating an irrigation system required for the city park's soccer fields. Which alternative should be selected, if the interest rate is 12% and there is no salvage value?

Year	A	B
0	−$15,000	−$25,000
1–10	+5,310	+7,900

Solution

One must be selected so only an incremental analysis is needed.

$\Delta Cost = -10,000$

$PW_{COST} = -10,000$

$\Delta Annual\ benefits = +2590$

$PW_{BENEFITS} = 2590(P/A, 12\%, 10)$

$\qquad\qquad = 14,634$

$\Delta B/C = 14,634/10,000 = 1.46$

$1.46 > 1$; therefore choose B, the higher-cost alternative.

16-13

An impoundment may be built for the city's water supply. Two different sites are technically, politically, socially, and financially feasible. The city council has asked for a benefit/cost analysis at a 6% interest rate.

Year	Rattlesnake Canyon	Blue Basin
0	−$15,000,000	−$27,000,000
1–75	+2,000,000	+3,000,000

Which site should you recommend?

Solution

	Rattlesnake Canyon	Blue Basin

$$\frac{PW_B}{PW_C} = \frac{2\times10^6(P/A, 6\%, 75)}{15\times10^6} \qquad \frac{3\times10^6(P/A, 6\%, 75)}{27\times10^6}$$

B/C ratio   =   2.19 >1 (OK)   1.83 > 1 (OK)

Incremental analysis:

Year	BB – RC
0	–12,000,000
1–75	+1,000,000

$$\text{B/C ratio} \quad = \quad \frac{1\times10^6(P/A, 6\%, 75)}{12\times10^6} \quad = 1.37 > 1$$

∴ Choose the higher-cost alternative, Blue Basin.

16-14

A city engineer may accept one of two bids for new park equipment. Using benefit/cost analysis at a 10% interest rate should one be selected? If yes, which one?

Computer	Cost	Annual Benefits	Salvage	Useful Life
A	$48,000	$13,000	$0	6 years
B	40,000	12,000	0	6 years

Solution

Evaluate B first since it costs less.

Alternative *B*

$PW_B = 12,000(P/A, 10\%, 6) = \$52,260$

$PW_C = 40,000$

B/C   = 52,260/40,000 = 1.3065

Alternative *A*

$PW_B = 13,000(P/A, 10\%, 6) = \$56,615$

$PW_C = 48,000$

B/C   = 56,615/48,000 = 1.179

Incremental analysis

$$\frac{\Delta B}{\Delta C} = \frac{56,516-52,260}{48,000-40,000} = \frac{4355}{8000} = 0.544$$

$\dfrac{\Delta B}{\Delta C} < 1.0$ ∴ Select the less costly alternative, *B*.

16-15

The hotel and conference center at Wicker Valley State Park may be expanded.

	Alternative *A*	Alternative *B*	Alternative *C*
Investment cost	$180,000	$100,000	$280,000
Annual operating costs	16,000	12,000	28,000
Annual benefits	53,000	35,000	77,000

Use a MARR of 10% and a 10-year life and complete a benefit/cost ratio analysis.

Solution

1. B/C ratios of individual alternatives.

$$\text{B/C}_A = \frac{\text{AW}_\text{B}}{\text{AW}_\text{C}} = \frac{53,000}{180,000(A/P,10\%,10)+16,000} = 1.17$$

$$\text{B/C}_B = \frac{\text{AW}_\text{B}}{\text{AW}_\text{C}} = \frac{35,000}{100,000(A/P,10\%,10)+12,000} = 1.24$$

$$\text{B/C}_C = \frac{\text{AW}_\text{B}}{\text{AW}_\text{C}} = \frac{77,000}{280,000(A/P,10\%,10)+28,000} = 1.05$$

$\therefore$ All are economically attractive.

2. Incremental B/C analysis

    A–B

    ΔBenefits = 53,000 – 35,000 = 18,000

    ΔCosts = (180,000 – 1000,000)(A/P, 10%, 10) + (16,000 – 12,000) = 17,016

    $\text{B/C} = \dfrac{18,000}{17,016} = 1.06$ (> 1) $\therefore$ Choose *A*.

    C–A

    ΔBenefits = 77,000 – 53,000 = 24,000

    ΔCosts = (280,000 – 180,000)(A/P, 10%, 10) + (28,000 – 16,000) = 28,270

    $\text{B/C} = \dfrac{24,000}{28,270} = 0.85$ (< 1) $\therefore$ Choose *A*.

Select Alternative *A*. If operating costs were considered as reducing the annual benefits, the B/C ratios would differ but *A* would still be the best.

16-16

At an interest rate of 12% and a horizon of 15 years use benefit/cost ratios to choose which design is preferred.

	A	B	C
Initial Investment	$12,000	$17,000	$18,000
Annual Savings	$3,500	$4,500	$9,000
Annual Costs	$1,000	$2,500	$6,500
Salvage Value	$5,000	$4,500	$11,000

	A	B	C	D	E
1	12%	interest rate			
2	15	horizon (years)			
3		**A**	**B**	**C**	**C - A**
4	First cost	$ 12,000	$ 17,000	$ 18,000	$ 6,000
5	Annual savings	$ 3,500	$ 4,500	$ 9,000	$ 5,500
6	Annual costs	$ 1,000	$ 2,500	$ 6,500	$ 5,500
7	Salvage value	$ 5,000	$ 4,500	$ 11,000	$ 6,000
8					
9	PW benefits	$ 23,838	$ 30,649	$ 61,298	$ 37,460
10	PW costs	$ 17,897	$ 33,205	$ 60,261	$ 42,364
11	B/C	1.33	0.92	1.02	0.88

Solution

The B/C ratio places only the PW of the savings in the numerator. B is not attractive. C is less attractive than A.

16-17

Use benefit/cost ratio analysis to determine which alternative should be selected; each has a 6-year useful life. Assume a 10% MARR.

	A	*B*	*C*	*D*
First cost	$880	$560	$700	$900
Annual benefit	240	130	110	250
Annual cost	80	20	0	40
Salvage value	300	200	440	110

Solution

A

$AW_{BENEFITS} = 240 + 300(A/F, 10\%, 6) = 278$

$AW_{COSTS} \quad = 880(A/P, 10\%, 6) + 80 = 282$

$\qquad\qquad B/C = 0.98 \ (\,< 1, \text{ so eliminate Alternative } A)$

B

$AW_{BENEFITS} = 130 + 200(A/F, 10\%, 6) = 156$

$AW_{COSTS} \quad = 560(A/P, 10\%, 6) + 20 = 149$

$\qquad\qquad B/C = 1.04$

C

$AW_{BENEFITS} = 110 + 440(A/F, 10\%, 6) = 167$

$AW_{COSTS} \quad = 700(A/P, 10\%, 6) + 0 = 160$

$\qquad\qquad B/C = 1.04$

$\underline{D}$

$AW_{BENEFITS} = 250 + 110(A/F, 10\%, 6) = 264$

$AW_{COSTS}\quad = 900(A/P, 10\%, 6) + 40 = 246$

$\qquad B/C = 1.07$

B is the least expensive acceptable alternative.

$\underline{C-B}$

$AW_{BENEFITS} = 167 - 156 = 11$

$AW_{COSTS}\quad = 160 - 149 = 11$

$B/C = 1.00 \quad \therefore \quad$ Choose C.

$\underline{D-C}$

$AW_{BENEFITS} = 264 - 167 = 97$

$AW_{COSTS}\quad = 246 - 160 = 86$

$\qquad B/C = 1.13 \quad \therefore \quad$ Choose D.

16-18

Four mutually exclusive alternatives, each with a 20-year life and no salvage value, have been presented to the city council of Anytown, U.S.A. Which alternative should be selected?

	A	B	C	D
PW of costs	$4000	$ 9,000	$6000	$2000
PW of benefits	6480	11,250	5700	4700

Solution

	A	B	C	D
B/C ratio =	1.62	2.35	0.95	2.35

$\qquad\qquad\qquad\qquad\qquad\qquad C < 1 \therefore$ eliminate

(rearrange by PW of costs)	D	A	B
PW of cost	$2000	$4000	$ 9,000
PW of benefits	4700	6480	11,250

$\underline{A-D}$

ΔB $1780

ΔC   2000 $\Delta B/\Delta C = 0.89 < 1.0 \therefore$ Choose D, the least expensive alternative.

$\underline{B-D}$

ΔB $6550

ΔC   7000 $\Delta B/\Delta C = 0.94 < 1.0 \therefore$ Choose D, the least expensive alternative.

16-19

The city council of Arson, Michigan, is debating whether to buy a new fire truck

	Truck A	Truck B
First cost	$50,000	$60,000
Annual maintenance	5,000	4,000
Salvage value	5,000	6,000
Annual reduction in fire damage	20,000	21,000
Useful life	7 years	7 years

a. Use the modified B/C ratio method to determine whether the city should buy a new truck, and if so, which one to buy, assuming that it will be paid for with money borrowed at an interest rate of 7%.

b. How would the decision be affected if inflation were considered? Assume that maintenance cost, salvage value, and fire damage are responsive to inflation.

Solution

a. In the modified B/C ratio, all annual cash flows are in the numerator, while first cost and salvage are in the denominator. Either present worth or uniform equivalent methods may be used.

$$B/C_A = \frac{20,000 - 5000}{50,000(A/P, 7\%, 7) - 5000(A/F, 7\%, 7)} = 1.72 \qquad (>1) \therefore A \text{ is acceptable}$$

$$B/C_B = \frac{21,000 - 4000}{60,000(A/P, 7\%, 7) - 6000(A/F, 7\%, 7)} = 1.63 \qquad (>1) \therefore B \text{ is acceptable}$$

$$B/C_{B-A} = \frac{1000 - (-1000)}{10,000(A/P, 7\%, 7) - 1000(A/F, 7\%, 7)} = 1.15 \qquad (>1) \therefore B \text{ is better than } A$$

Truck B should be purchased.

b. Since both future costs (maintenance) and benefits (reduced damage and salvage) are responsive to inflation, the decision is not affected by inflation.

Chapter 17

Accounting and Engineering Economy

17-1

Determine retained earnings for Lavelle Manufacturing.

Current liabilities	$4,000,000
Long-term liabilities	2,000,000
Current assets	6,500,000
Fixed assets	4,000,000
Other assets	1,500,000
Common stock	2,500,000
Preferred stock	500,000
Capital surplus	1,000,000

Solution

Equity = Common stock + Preferred stock + Capital surplus + Retained earnings

Equity = 2.5M + 0.5M + 1M + Retained earnings

$\qquad$ = 4M + Retained earnings

Assets = Liabilities + Equity

6.5M + 4M + 1.5M = (4M + 2M) + (4M + Retained earnings)

Retained earnings = $2,000,000

17-2

Billy Bob's Towing and Repair Service has provided the following financial information.

Cash	$ 80,000
Accounts receivable	120,000
Accounts payable	200,000
Securities	75,000
Parts inventories	42,000
Prepaid expenses	30,000
Accrued expense	15,000

Determine (a) the current ratio, (b) the quick ratio, and (c) the available working capital.

Solution

a.   Current ratio $= \dfrac{\text{Current Assets}}{\text{Current Liabilities}} = \dfrac{80{,}000 + 120{,}000 + 75{,}000 + 42{,}000}{200{,}000 + 15{,}000} = 1.47$

b.   Quick ratio $= \dfrac{\text{Current Assets - Inventories}}{\text{Current Liabilities}} = \dfrac{80{,}000 + 120{,}000 + 75{,}000}{200{,}000 + 15{,}000} = 1.28$

c.   Working capital = Current assets – Current liabilities = \$102,000

17-3

Determine the current and quick ratios for Harbor Master Boats Inc. Does the firm appear to be financially sound?

Harbor Master Boats Inc.
Balance Sheet, January 1, 20XX

Assets		Liabilities	
Current assets		Current liabilities	
Cash	900,000	Accounts payable	2,400,000
Accounts receivable	1,100,000	Notes payable	2,000,000
Inventory	2,000,000	Accrued expense	900,000
Total current assets	4,000,000	Total current liabilities	5,300,000
Fixed assets		Long-term debt	3,000,000
Land	300,000	Total liabilities	8,300,000
Plant	2,500,000	**Equity**	
Equipment	6,000,000	Stock	2,000,000
Total fixed assets	8,800,000	Retained earnings	2,500,000
		Total net worth	4,500,000
Total assets	12,800,000	Total liabilities and net worth	12,800,000

Solution

Current ratio $= \dfrac{\text{Current Assets}}{\text{Current Liabilities}} = \dfrac{4{,}000{,}000}{5{,}300{,}000} = .755$

Quick ratio $= \dfrac{\text{Current Assets - Inventories}}{\text{Current Liabilities}} = \dfrac{4{,}000{,}000 - 2{,}000{,}000}{5{,}300{,}000} = .377$

Based on these ratios, the firm is not very sound financially. The current ratio typically should be greater than 2. The quick ratio indicates the company's inability to pay off current liabilities with "quick" capital.

17-4

Rapid Delivery's financial information includes:

Acid-test ratio	1.3867
Cash on hand	$ 72,000
Accounts receivable	102,000
Market value of securities held	34,000
Inventories	143,000
Other assets	16,000
Fixed assets	215,000
Total liabilities	400,000

Determine (a) the current assets, (b) the current liabilities, (c) the total assets, and (d) the owner's equity.

Solution

a. Current assets = 72,000 + 102,000 + 34,000 + 143,000 = $351,000

b. Acid-test ratio $= \dfrac{\text{Current Assets – Inventories}}{\text{Current Liabilities}}$

Current liabilities $= \dfrac{\text{Current Assets – Inventories}}{\text{Acid-test ratio}} = \dfrac{351,000 - 143,000}{1.3867} = \$149,996$

c. Total assets = 351,000 + 215,000 + 16,000 = $582,000

d. Owner's equity = Total assets – Total liabilities
$$= 582,000 - 400,000$$
$$= \$182,000$$

17-5

The following information has been taken from the financial statements available for the ABC Company.

Accounts payable	$ 4,000
Accounts receivable	12,000
Income taxes	6,000
Owner's equity	75,000
Cost of goods sold	42,000
Selling expense	10,000
Sales revenue	80,000

Determine the net income.

Solution

Net income = Sales revenue – Cost of goods sold – Selling expense – Income taxes
$$= 80,000 - 42,000 - 10,000 - 6000$$
$$= \$22,000$$

17-6

This financial information is from Firerock Industries' income statement:

Revenues	
Sales	$3,200,000
Operating revenue	2,000,000
Nonoperating revenue	3,400,000
Expenses	
Total operating expenses	6,700,000
Interest payments	500,000

Taxes paid for the year equaled $110,000. Determine (a) the net income before taxes, (b) the net profit (loss), (c) the interest coverage, and (d) the net profit ratio.

Solution

a. Net income before taxes = 3.2M + 2.0M + 3.4M − 6.7M − 0.5M = $1,400,000

b. Net profit = 1.4M − 0.11M = $1,290,000

c. Interest coverage = $\dfrac{\text{Total Income}}{\text{Interest Payments}} = \dfrac{1.9M}{0.5M} = 3.8$

d. Net profit ratio = $\dfrac{\text{Net Profit}}{\text{Net Sales Revenue}} = \dfrac{1.29M}{3.2M} = 0.4031$

17-7

Wheeler Industries had sales of $157M and returns and allowances of $21M. Operating expenses were $48M and nonoperating revenues and expenses equaled the interest paid of $7M. Using the corporate tax rate from the TCJA compute the tax paid by Wheeler Industries. Construct the income statement.

Solution

	A	B	C	D
1	Operating revenues and expenses	(in $M)		
2	Operating revenues			
3	Sales	$157		
4	(minus) Returns and allowances	-$21		
5	Total operating revenues	$136		
6	Total operating expenses	$48		
7	Total operating income	$88		
8				
9	Total non-operating income	-$7		
10				
11	Net income before taxes	$81		
12	Income taxes	$17.01	21%	tax rate
13	Net profit (loss)	$64		

17-8

Wheeler Industries (problem SG17-7) had the following entries in its balance sheet at the end of last year.

Plant and equipment	$17M
(less accumulated depreciation)	9M
Retained earnings	$80M

In addition to the income statement data for this year in problem SG17-7, we also know that the firm purchased $5 million of equipment with cash and that depreciation expenses were $4 million of the $48 million in operating expenses. The firm paid no dividends and sold no stock this year. What are the entries in the balance sheet at the end of this year for:

(a) Plant and equipment?

(b) Accumulated depreciation?

(c) Retained earnings?

Solution

(a) Plant and equipment = $17M + $5M = $22M

(b) Accumulated depreciation = $9M + $4M = $13M

(c) $RE_{end} = RE_{begin}$ + Net income or Loss (including taxes) + New Stock − Dividends

   = $80M + $64M (from SG17-7) + 0 − 0 = $144M

17-9

Brown Box Inc., manufactures shipping boxes for a wide variety of industries. Their XLLarge costs:

Direct materials costs	$0.85/unit
Direct labor costs	3.85/unit

Overhead for the entire manufacturing plant is $4,000,000 per year, and it is allocated based on direct labor costs. Total direct labor costs are $5,500,000. The demand for XLLarge is 200,000 boxes per year. Determine the cost per unit.

Solution

Cost per unit = Direct materials cost + Direct labor costs + Overhead costs
Overhead cost allocation:

Direct labor cost = 200,000 × $3.85 = $770,000

Allocation of overhead = $4,000,000 \times \frac{770,000}{5,500,000}$ = 400,000/200,000 = $2.80/box

Cost per unit = 0.85 + 3.85 + 2.80 = $7.50

17-10

Abby Manufacturing produces children's toys and the Dr. Dolittle Farm is a big seller. Indirect costs of $750,000 are allocated based on direct materials cost, which total $8,350,000 for the facility. The materials used in Dr. Dolittle Farm cost $7.45 per unit. The total labor (both direct and indirect) costs $9.35 per unit. The production schedule is 300,000 units for the coming year. If Abby desires a 35% profit margin, what should the Farm's wholesale price be?

Solution

Total labor $= \$9.35 \times 300,000 = \$2,805,000$

Total materials $= \$7.45 \times 300,000 = \$2,235,000$

Overhead $= (\$2,235,000/\$8,350,000) \times \$750,000 = \$200,750$

Total production cost $= \$2,805,000 + \$2,235,000 + \$200,750 = \$5,240,750$

Cost per unit $= \$5,240,750/300,000 = \$17.47$

Wholesale price $= \$17.47 \times 1.35 = \$23.58$

17-11

Denali Industries makes specialized oil well equipment. Use total direct cost as the burden vehicle, and compute the total cost per unit for each model. Total manufacturing indirect costs are $2,500,000, and there are 1000 units manufactured per year for Model A, 600 for Model K, and 500 for Model T. Use total direct cost as the burden vehicle and compute the total cost per unit for each model.

Activity	Model A	Model K	Model T
Direct material cost	$800,000	$530,000	$630,000
Direct labor cost	$500,000	$320,000	$410,000

Solution

	A	B	C	D	E
1	**Activity**	**Model A**	**Model K**	**Model T**	**Total**
2	Direct material cost	$800,000	$530,000	$630,000	$1,960,000
3	Direct labor cost	$500,000	$320,000	$410,000	$1,230,000
4	Total direct cost	$1,300,000	$850,000	$1,040,000	$3,190,000
5	% total direct	41%	27%	33%	
6	Allocated indirect cost	$1,018,809	$666,144	$815,047	$2,500,000
7	Units produced	1,000	600	500	2,100
8	Cost per unit	$2,319	$2,527	$3,710	

Appendix B

Time Value of Money Calculations Using Spreadsheets and Calculators

B-1

Installing a heat exchanger in an office building will cost $750,000. It will save $65,000 annually for 20 years, when it will have a salvage value of $40,000. What is the PW of the heat exchanger at an interest rate of 5%?

Solution

	A	B	C	D	E	F	G	H	I
1	*i*	*n*	*PMT*	*PV*	*FV*	Solve for	Answer		
2	5.0%	20	65,000		40,000	PV	-$825,119	=PV(A2,B2,C2,E2)	
3							$825,119	change sign	
4	PMT, PV, & FV reverse sign of unknown						-$750,000	subtract first cost	
5	User must decide what sign is for next step						$75,119	PW	

B-2

Suppose that a computer costs you $1500 to buy and you sell it 3 years later for $300. If your interest rate for the time value of money is 8%, what is your annual cost of ownership?

Solution

	A	B	C	D	E	F	G	H	I
1	*i*	*n*	*PMT*	*PV*	*FV*	Solve for	Answer		
2	8.0%	3		-1500	300	PMT	$490	=PMT(A2,B2,D2,E2)	
3								= annual cost	
4	PMT, PV, & FV reverse sign of unknown								
5	User must decide what sign is for next step								

B-3

How many months to pay off a credit card balance of $8000 with payments of $200? The card's monthly interest rate is 1.5%. Assume there are no new charges.

Solution

When the credit card is paid off, the final or future value is 0.

	A	B	C	D	E	F	G	H	I
1	*i*	*n*	*PMT*	*PV*	*FV*	Solve for	Answer		
2	1.5%		200	-8000	0	NPER	61.5	=NPER(A2,C2,D2,E2)	
3								= # months	

COMPOUND INTEREST TABLES

Values of Interest Factors When n Equals Infinity

Single Payment:

$(F/P, i, \infty) = \infty$

$(P/F, i, \infty) = 0$

Arithmetic Gradient Series:

$(A/G, i, \infty) = 1/i$

$(P/G, i, \infty) = 1/i^2$

Uniform Payment Series:

$(A/F, i, \infty) = 0$

$(A/P, i, \infty) = i$

$(F/A, i, \infty) = \infty$

$(P/A, i, \infty) = 1/i$

1/4%
Compound Interest Factors
1/4%

	Single Payment		Uniform Payment Series				Arithmetic Gradient		
	Compound Amount Factor	Present Worth Factor	Sinking Fund Factor	Capital Recovery Factor	Compound Amount Factor	Present Worth Factor	Gradient Uniform Series	Gradient Present Worth	
	Find F Given P	Find P Given F	Find A Given F	Find A Given P	Find F Given A	Find P Given A	Find A Given G	Find P Given G	
n	F/P	P/F	A/F	A/P	F/A	P/A	A/G	P/G	n
1	1.003	.9975	1.0000	1.0025	1.000	0.998	0.000	0.000	1
2	1.005	.9950	.4994	.5019	2.003	1.993	0.499	0.995	2
3	1.008	.9925	.3325	.3350	3.008	2.985	0.998	2.980	3
4	1.010	.9901	.2491	.2516	4.015	3.975	1.497	5.950	4
5	1.013	.9876	.1990	.2015	5.025	4.963	1.995	9.901	5
6	1.015	.9851	.1656	.1681	6.038	5.948	2.493	14.826	6
7	1.018	.9827	.1418	.1443	7.053	6.931	2.990	20.722	7
8	1.020	.9802	.1239	.1264	8.070	7.911	3.487	27.584	8
9	1.023	.9778	.1100	.1125	9.091	8.889	3.983	35.406	9
10	1.025	.9753	.0989	.1014	10.113	9.864	4.479	44.184	10
11	1.028	.9729	.0898	.0923	11.139	10.837	4.975	53.913	11
12	1.030	.9705	.0822	.0847	12.167	11.807	5.470	64.589	12
13	1.033	.9681	.0758	.0783	13.197	12.775	5.965	76.205	13
14	1.036	.9656	.0703	.0728	14.230	13.741	6.459	88.759	14
15	1.038	.9632	.0655	.0680	15.266	14.704	6.953	102.244	15
16	1.041	.9608	.0613	.0638	16.304	15.665	7.447	116.657	16
17	1.043	.9584	.0577	.0602	17.344	16.624	7.944	131.992	17
18	1.046	.9561	.0544	.0569	18.388	17.580	8.433	148.245	18
19	1.049	.9537	.0515	.0540	19.434	18.533	8.925	165.411	19
20	1.051	.9513	.0488	.0513	20.482	19.485	9.417	183.485	20
21	1.054	.9489	.0464	.0489	21.534	20.434	9.908	202.463	21
22	1.056	.9465	.0443	.0468	22.587	21.380	10.400	222.341	22
23	1.059	.9442	.0423	.0448	23.644	22.324	10.890	243.113	23
24	1.062	.9418	.0405	.0430	24.703	23.266	11.380	264.775	24
25	1.064	.9395	.0388	.0413	25.765	24.206	11.870	287.323	25
26	1.067	.9371	.0373	.0398	26.829	25.143	12.360	310.752	26
27	1.070	.9348	.0358	.0383	27.896	26.078	12.849	335.057	27
28	1.072	.9325	.0345	.0370	28.966	27.010	13.337	360.233	28
29	1.075	.9301	.0333	.0358	30.038	27.940	13.825	386.278	29
30	1.078	.9278	.0321	.0346	31.114	28.868	14.313	413.185	30
36	1.094	.9140	.0266	.0291	37.621	34.387	17.231	592.499	36
40	1.105	.9049	.0238	.0263	42.014	38.020	19.167	728.740	40
48	1.127	.8871	.0196	.0221	50.932	45.179	23.021	1 040.055	48
50	1.133	.8826	.0188	.0213	53.189	46.947	23.980	1 125.777	50
52	1.139	.8782	.0180	.0205	55.458	48.705	24.938	1 214.588	52
60	1.162	.8609	.0155	.0180	64.647	55.653	28.751	1 600.085	60
70	1.191	.8396	.0131	.0156	76.395	64.144	33.481	2 147.611	70
72	1.197	.8355	.0127	.0152	78.780	65.817	34.422	2 265.557	72
80	1.221	.8189	.0113	.0138	88.440	72.427	38.169	2 764.457	80
84	1.233	.8108	.0107	.0132	93.343	75.682	40.033	3 029.759	84
90	1.252	.7987	.00992	.0124	100.789	80.504	42.816	3 446.870	90
96	1.271	.7869	.00923	.0117	108.349	85.255	45.584	3 886.283	96
100	1.284	.7790	.00881	.0113	113.451	88.383	47.422	4 191.242	100
104	1.297	.7713	.00843	.0109	118.605	91.480	49.252	4 505.557	104
120	1.349	.7411	.00716	.00966	139.743	103.563	56.508	5 852.112	120
240	1.821	.5492	.00305	.00555	328.306	180.312	107.586	19 398.985	240
360	2.457	.4070	.00172	.00422	582.745	237.191	152.890	36 263.930	360
480	3.315	.3016	.00108	.00358	926.074	279.343	192.670	53 820.752	480

1/2%

Compound Interest Factors

1/2%

	Single Payment		Uniform Payment Series				Arithmetic Gradient		
	Compound Amount Factor	Present Worth Factor	Sinking Fund Factor	Capital Recovery Factor	Compound Amount Factor	Present Worth Factor	Gradient Uniform Series	Gradient Present Worth	
	Find F Given P F/P	Find P Given F P/F	Find A Given F A/F	Find A Given P A/P	Find F Given A F/A	Find P Given A P/A	Find A Given G A/G	Find P Given G P/G	
n									n
1	1.005	.9950	1.0000	1.0050	1.000	0.995	0	0	1
2	1.010	.9901	.4988	.5038	2.005	1.985	0.499	0.991	2
3	1.015	.9851	.3317	.3367	3.015	2.970	0.996	2.959	3
4	1.020	.9802	.2481	.2531	4.030	3.951	1.494	5.903	4
5	1.025	.9754	.1980	.2030	5.050	4.926	1.990	9.803	5
6	1.030	.9705	.1646	.1696	6.076	5.896	2.486	14.660	6
7	1.036	.9657	.1407	.1457	7.106	6.862	2.980	20.448	7
8	1.041	.9609	.1228	.1278	8.141	7.823	3.474	27.178	8
9	1.046	.9561	.1089	.1139	9.182	8.779	3.967	34.825	9
10	1.051	.9513	.0978	.1028	10.228	9.730	4.459	43.389	10
11	1.056	.9466	.0887	.0937	11.279	10.677	4.950	52.855	11
12	1.062	.9419	.0811	.0861	12.336	11.619	5.441	63.218	12
13	1.067	.9372	.0746	.0796	13.397	12.556	5.931	74.465	13
14	1.072	.9326	.0691	.0741	14.464	13.489	6.419	86.590	14
15	1.078	.9279	.0644	.0694	15.537	14.417	6.907	99.574	15
16	1.083	.9233	.0602	.0652	16.614	15.340	7.394	113.427	16
17	1.088	.9187	.0565	.0615	17.697	16.259	7.880	128.125	17
18	1.094	.9141	.0532	.0582	18.786	17.173	8.366	143.668	18
19	1.099	.9096	.0503	.0553	19.880	18.082	8.850	160.037	19
20	1.105	.9051	.0477	.0527	20.979	18.987	9.334	177.237	20
21	1.110	.9006	.0453	.0503	22.084	19.888	9.817	195.245	21
22	1.116	.8961	.0431	.0481	23.194	20.784	10.300	214.070	22
23	1.122	.8916	.0411	.0461	24.310	21.676	10.781	233.680	23
24	1.127	.8872	.0393	.0443	25.432	22.563	11.261	254.088	24
25	1.133	.8828	.0377	.0427	26.559	23.446	11.741	275.273	25
26	1.138	.8784	.0361	.0411	27.692	24.324	12.220	297.233	26
27	1.144	.8740	.0347	.0397	28.830	25.198	12.698	319.955	27
28	1.150	.8697	.0334	.0384	29.975	26.068	13.175	343.439	28
29	1.156	.8653	.0321	.0371	31.124	26.933	13.651	367.672	29
30	1.161	.8610	.0310	.0360	32.280	27.794	14.127	392.640	30
36	1.197	.8356	.0254	.0304	39.336	32.871	16.962	557.564	36
40	1.221	.8191	.0226	.0276	44.159	36.172	18.836	681.341	40
48	1.270	.7871	.0185	.0235	54.098	42.580	22.544	959.928	48
50	1.283	.7793	.0177	.0227	56.645	44.143	23.463	1 035.70	50
52	1.296	.7716	.0169	.0219	59.218	45.690	24.378	1 113.82	52
60	1.349	.7414	.0143	.0193	69.770	51.726	28.007	1 448.65	60
70	1.418	.7053	.0120	.0170	83.566	58.939	32.468	1 913.65	70
72	1.432	.6983	.0116	.0166	86.409	60.340	33.351	2 012.35	72
80	1.490	.6710	.0102	.0152	98.068	65.802	36.848	2 424.65	80
84	1.520	.6577	.00961	.0146	104.074	68.453	38.576	2 640.67	84
90	1.567	.6383	.00883	.0138	113.311	72.331	41.145	2 976.08	90
96	1.614	.6195	.00814	.0131	122.829	76.095	43.685	3 324.19	96
100	1.647	.6073	.00773	.0127	129.334	78.543	45.361	3 562.80	100
104	1.680	.5953	.00735	.0124	135.970	80.942	47.025	3 806.29	104
120	1.819	.5496	.00610	.0111	163.880	90.074	53.551	4 823.52	120
240	3.310	.3021	.00216	.00716	462.041	139.581	96.113	13 415.56	240
360	6.023	.1660	.00100	.00600	1 004.5	166.792	128.324	21 403.32	360
480	10.957	.0913	.00050	.00550	1 991.5	181.748	151.795	27 588.37	480

3/4%

Compound Interest Factors

3/4%

	Single Payment		Uniform Payment Series				Arithmetic Gradient		
	Compound Amount Factor	Present Worth Factor	Sinking Fund Factor	Capital Recovery Factor	Compound Amount Factor	Present Worth Factor	Gradient Uniform Series	Gradient Present Worth	
	Find F Given P	Find P Given F	Find A Given F	Find A Given P	Find F Given A	Find P Given A	Find A Given G	Find P Given G	
n	F/P	P/F	A/F	A/P	F/A	P/A	A/G	P/G	n
1	1.008	.9926	1.0000	1.0075	1.000	0.993	0	0	1
2	1.015	.9852	.4981	.5056	2.008	1.978	0.499	0.987	2
3	1.023	.9778	.3308	.3383	3.023	2.956	0.996	2.943	3
4	1.030	.9706	.2472	.2547	4.045	3.926	1.492	5.857	4
5	1.038	.9633	.1970	.2045	5.076	4.889	1.986	9.712	5
6	1.046	.9562	.1636	.1711	6.114	5.846	2.479	14.494	6
7	1.054	.9490	.1397	.1472	7.160	6.795	2.971	20.187	7
8	1.062	.9420	.1218	.1293	8.213	7.737	3.462	26.785	8
9	1.070	.9350	.1078	.1153	9.275	8.672	3.951	34.265	9
10	1.078	.9280	.0967	.1042	10.344	9.600	4.440	42.619	10
11	1.086	.9211	.0876	.0951	11.422	10.521	4.927	51.831	11
12	1.094	.9142	.0800	.0875	12.508	11.435	5.412	61.889	12
13	1.102	.9074	.0735	.0810	13.602	12.342	5.897	72.779	13
14	1.110	.9007	.0680	.0755	14.704	13.243	6.380	84.491	14
15	1.119	.8940	.0632	.0707	15.814	14.137	6.862	97.005	15
16	1.127	.8873	.0591	.0666	16.932	15.024	7.343	110.318	16
17	1.135	.8807	.0554	.0629	18.059	15.905	7.822	124.410	17
18	1.144	.8742	.0521	.0596	19.195	16.779	8.300	139.273	18
19	1.153	.8676	.0492	.0567	20.339	17.647	8.777	154.891	19
20	1.161	.8612	.0465	.0540	21.491	18.508	9.253	171.254	20
21	1.170	.8548	.0441	.0516	22.653	19.363	9.727	188.352	21
22	1.179	.8484	.0420	.0495	23.823	20.211	10.201	206.170	22
23	1.188	.8421	.0400	.0475	25.001	21.053	10.673	224.695	23
24	1.196	.8358	.0382	.0457	26.189	21.889	11.143	243.924	24
25	1.205	.8296	.0365	.0440	27.385	22.719	11.613	263.834	25
26	1.214	.8234	.0350	.0425	28.591	23.542	12.081	284.421	26
27	1.224	.8173	.0336	.0411	29.805	24.360	12.548	305.672	27
28	1.233	.8112	.0322	.0397	31.029	25.171	13.014	327.576	28
29	1.242	.8052	.0310	.0385	32.261	25.976	13.479	350.122	29
30	1.251	.7992	.0298	.0373	33.503	26.775	13.942	373.302	30
36	1.309	.7641	.0243	.0318	41.153	31.447	16.696	525.038	36
40	1.348	.7416	.0215	.0290	46.447	34.447	18.507	637.519	40
48	1.431	.6986	.0174	.0249	57.521	40.185	22.070	886.899	48
50	1.453	.6882	.0166	.0241	60.395	41.567	22.949	953.911	50
52	1.475	.6780	.0158	.0233	63.312	42.928	23.822	1 022.64	52
60	1.566	.6387	.0133	.0208	75.425	48.174	27.268	1 313.59	60
70	1.687	.5927	.0109	.0184	91.621	54.305	31.465	1 708.68	70
72	1.713	.5839	.0105	.0180	95.008	55.477	32.289	1 791.33	72
80	1.818	.5500	.00917	.0167	109.074	59.995	35.540	2 132.23	80
84	1.873	.5338	.00859	.0161	116.428	62.154	37.137	2 308.22	84
90	1.959	.5104	.00782	.0153	127.881	65.275	39.496	2 578.09	90
96	2.049	.4881	.00715	.0147	139.858	68.259	41.812	2 854.04	96
100	2.111	.4737	.00675	.0143	148.147	70.175	43.332	3 040.85	100
104	2.175	.4597	.00638	.0139	156.687	72.035	44.834	3 229.60	104
120	2.451	.4079	.00517	.0127	193.517	78.942	50.653	3 998.68	120
240	6.009	.1664	.00150	.00900	667.901	111.145	85.422	9 494.26	240
360	14.731	.0679	.00055	.00805	1 830.8	124.282	107.115	13 312.50	360
480	36.111	.0277	.00021	.00771	4 681.5	129.641	119.662	15 513.16	480

1%   Compound Interest Factors   1%

	Single Payment		Uniform Payment Series				Arithmetic Gradient		
	Compound Amount Factor	Present Worth Factor	Sinking Fund Factor	Capital Recovery Factor	Compound Amount Factor	Present Worth Factor	Gradient Uniform Series	Gradient Present Worth	
	Find F Given P	Find P Given F	Find A Given F	Find A Given P	Find F Given A	Find P Given A	Find A Given G	Find P Given G	
n	F/P	P/F	A/F	A/P	F/A	P/A	A/G	P/G	n
1	1.010	.9901	1.0000	1.0100	1.000	0.990	0	0	1
2	1.020	.9803	.4975	.5075	2.010	1.970	0.498	0.980	2
3	1.030	.9706	.3300	.3400	3.030	2.941	0.993	2.921	3
4	1.041	.9610	.2463	.2563	4.060	3.902	1.488	5.804	4
5	1.051	.9515	.1960	.2060	5.101	4.853	1.980	9.610	5
6	1.062	.9420	.1625	.1725	6.152	5.795	2.471	14.320	6
7	1.072	.9327	.1386	.1486	7.214	6.728	2.960	19.917	7
8	1.083	.9235	.1207	.1307	8.286	7.652	3.448	26.381	8
9	1.094	.9143	.1067	.1167	9.369	8.566	3.934	33.695	9
10	1.105	.9053	.0956	.1056	10.462	9.471	4.418	41.843	10
11	1.116	.8963	.0865	.0965	11.567	10.368	4.900	50.806	11
12	1.127	.8874	.0788	.0888	12.682	11.255	5.381	60.568	12
13	1.138	.8787	.0724	.0824	13.809	12.134	5.861	71.112	13
14	1.149	.8700	.0669	.0769	14.947	13.004	6.338	82.422	14
15	1.161	.8613	.0621	.0721	16.097	13.865	6.814	94.481	15
16	1.173	.8528	.0579	.0679	17.258	14.718	7.289	107.273	16
17	1.184	.8444	.0543	.0643	18.430	15.562	7.761	120.783	17
18	1.196	.8360	.0510	.0610	19.615	16.398	8.232	134.995	18
19	1.208	.8277	.0481	.0581	20.811	17.226	8.702	149.895	19
20	1.220	.8195	.0454	.0554	22.019	18.046	9.169	165.465	20
21	1.232	.8114	.0430	.0530	23.239	18.857	9.635	181.694	21
22	1.245	.8034	.0409	.0509	24.472	19.660	10.100	198.565	22
23	1.257	.7954	.0389	.0489	25.716	20.456	10.563	216.065	23
24	1.270	.7876	.0371	.0471	26.973	21.243	11.024	234.179	24
25	1.282	.7798	.0354	.0454	28.243	22.023	11.483	252.892	25
26	1.295	.7720	.0339	.0439	29.526	22.795	11.941	272.195	26
27	1.308	.7644	.0324	.0424	30.821	23.560	12.397	292.069	27
28	1.321	.7568	.0311	.0411	32.129	24.316	12.852	312.504	28
29	1.335	.7493	.0299	.0399	33.450	25.066	13.304	333.486	29
30	1.348	.7419	.0287	.0387	34.785	25.808	13.756	355.001	30
36	1.431	.6989	.0232	.0332	43.077	30.107	16.428	494.620	36
40	1.489	.6717	.0205	.0305	48.886	32.835	18.178	596.854	40
48	1.612	.6203	.0163	.0263	61.223	37.974	21.598	820.144	48
50	1.645	.6080	.0155	.0255	64.463	39.196	22.436	879.417	50
52	1.678	.5961	.0148	.0248	67.769	40.394	23.269	939.916	52
60	1.817	.5504	.0122	.0222	81.670	44.955	26.533	1 192.80	60
70	2.007	.4983	.00993	.0199	100.676	50.168	30.470	1 528.64	70
72	2.047	.4885	.00955	.0196	104.710	51.150	31.239	1 597.86	72
80	2.217	.4511	.00822	.0182	121.671	54.888	34.249	1 879.87	80
84	2.307	.4335	.00765	.0177	130.672	56.648	35.717	2 023.31	84
90	2.449	.4084	.00690	.0169	144.863	59.161	37.872	2 240.56	90
96	2.599	.3847	.00625	.0163	159.927	61.528	39.973	2 459.42	96
100	2.705	.3697	.00587	.0159	170.481	63.029	41.343	2 605.77	100
104	2.815	.3553	.00551	.0155	181.464	64.471	42.688	2 752.17	104
120	3.300	.3030	.00435	.0143	230.039	69.701	47.835	3 334.11	120
240	10.893	.0918	.00101	.0110	989.254	90.819	75.739	6 878.59	240
360	35.950	.0278	.00029	.0103	3 495.0	97.218	89.699	8 720.43	360
480	118.648	.00843	.00008	.0101	11 764.8	99.157	95.920	9 511.15	480

1 1/4%

Compound Interest Factors

1 1/4%

	Single Payment		Uniform Payment Series				Arithmetic Gradient		
	Compound Amount Factor	Present Worth Factor	Sinking Fund Factor	Capital Recovery Factor	Compound Amount Factor	Present Worth Factor	Gradient Uniform Series	Gradient Present Worth	
	Find F Given P	Find P Given F	Find A Given F	Find A Given P	Find F Given A	Find P Given A	Find A Given G	Find P Given G	
n	F/P	P/F	A/F	A/P	F/A	P/A	A/G	P/G	n
1	1.013	.9877	1.0000	1.0125	1.000	0.988	0	0	1
2	1.025	.9755	.4969	.5094	2.013	1.963	0.497	0.976	2
3	1.038	.9634	.3292	.3417	3.038	2.927	0.992	2.904	3
4	1.051	.9515	.2454	.2579	4.076	3.878	1.485	5.759	4
5	1.064	.9398	.1951	.2076	5.127	4.818	1.976	9.518	5
6	1.077	.9282	.1615	.1740	6.191	5.746	2.464	14.160	6
7	1.091	.9167	.1376	.1501	7.268	6.663	2.951	19.660	7
8	1.104	.9054	.1196	.1321	8.359	7.568	3.435	25.998	8
9	1.118	.8942	.1057	.1182	9.463	8.462	3.918	33.152	9
10	1.132	.8832	.0945	.1070	10.582	9.346	4.398	41.101	10
11	1.146	.8723	.0854	.0979	11.714	10.218	4.876	49.825	11
12	1.161	.8615	.0778	.0903	12.860	11.079	5.352	59.302	12
13	1.175	.8509	.0713	.0838	14.021	11.930	5.827	69.513	13
14	1.190	.8404	.0658	.0783	15.196	12.771	6.299	80.438	14
15	1.205	.8300	.0610	.0735	16.386	13.601	6.769	92.058	15
16	1.220	.8197	.0568	.0693	17.591	14.420	7.237	104.355	16
17	1.235	.8096	.0532	.0657	18.811	15.230	7.702	117.309	17
18	1.251	.7996	.0499	.0624	20.046	16.030	8.166	130.903	18
19	1.266	.7898	.0470	.0595	21.297	16.849	8.628	145.119	19
20	1.282	.7800	.0443	.0568	22.563	17.599	9.088	159.940	20
21	1.298	.7704	.0419	.0544	23.845	18.370	9.545	175.348	21
22	1.314	.7609	.0398	.0523	25.143	19.131	10.001	191.327	22
23	1.331	.7515	.0378	.0503	26.458	19.882	10.455	207.859	23
24	1.347	.7422	.0360	.0485	27.788	20.624	10.906	224.930	24
25	1.364	.7330	.0343	.0468	29.136	21.357	11.355	242.523	25
26	1.381	.7240	.0328	.0453	30.500	22.081	11.803	260.623	26
27	1.399	.7150	.0314	.0439	31.881	22.796	12.248	279.215	27
28	1.416	.7062	.0300	.0425	32.280	23.503	12.691	298.284	28
29	1.434	.6975	.0288	.0413	34.696	24.200	13.133	317.814	29
30	1.452	.6889	.0277	.0402	36.129	24.889	13.572	337.792	30
36	1.564	.6394	.0222	.0347	45.116	28.847	16.164	466.297	36
40	1.644	.6084	.0194	.0319	51.490	31.327	17.852	559.247	40
48	1.845	.5509	.0153	.0278	65.229	35.932	21.130	759.248	48
50	1.861	.5373	.0145	.0270	68.882	37.013	21.930	811.692	50
52	1.908	.5242	.0138	.0263	72.628	38.068	22.722	864.960	52
60	2.107	.4746	.0113	.0238	88.575	42.035	25.809	1 084.86	60
70	2.386	.4191	.00902	.0215	110.873	46.470	29.492	1 370.47	70
72	2.446	.4088	.00864	.0211	115.675	47.293	30.205	1 428.48	72
80	2.701	.3702	.00735	.0198	136.120	50.387	32.983	1 661.89	80
84	2.839	.3522	.00680	.0193	147.130	51.822	34.326	1 778.86	84
90	3.059	.3269	.00607	.0186	164.706	53.846	36.286	1 953.85	90
96	3.296	.3034	.00545	.0179	183.643	55.725	38.180	2 127.55	96
100	3.463	.2887	.00507	.0176	197.074	56.901	39.406	2 242.26	100
104	3.640	.2747	.00474	.0172	211.190	58.021	40.604	2 355.90	104
120	4.440	.2252	.00363	.0161	275.220	61.983	45.119	2 796.59	120
240	19.716	.0507	.00067	.0132	1 497.3	75.942	67.177	5 101.55	240
360	87.543	.0114	.00014	.0126	6 923.4	79.086	75.840	5 997.91	360
480	388.713	.00257	.00003	.0125	31 017.1	79.794	78.762	6 284.74	480

1 1/2 %

Compound Interest Factors

1 1/2 %

	Single Payment		Uniform Payment Series				Arithmetic Gradient		
	Compound Amount Factor	Present Worth Factor	Sinking Fund Factor	Capital Recovery Factor	Compound Amount Factor	Present Worth Factor	Gradient Uniform Series	Gradient Present Worth	
	Find F Given P	Find P Given F	Find A Given F	Find A Given P	Find F Given A	Find P Given A	Find A Given G	Find P Given G	
n	F/P	P/F	A/F	A/P	F/A	P/A	A/G	P/G	n
1	1.015	.9852	1.0000	1.0150	1.000	0.985	0	0	1
2	1.030	.9707	.4963	.5113	2.015	1.956	0.496	0.970	2
3	1.046	.9563	.3284	.3434	3.045	2.912	0.990	2.883	3
4	1.061	.9422	.2444	.2594	4.091	3.854	1.481	5.709	4
5	1.077	.9283	.1941	.2091	5.152	4.783	1.970	9.422	5
6	1.093	.9145	.1605	.1755	6.230	5.697	2.456	13.994	6
7	1.110	.9010	.1366	.1516	7.323	6.598	2.940	19.400	7
8	1.126	.8877	.1186	.1336	8.433	7.486	3.422	25.614	8
9	1.143	.8746	.1046	.1196	9.559	8.360	3.901	32.610	9
10	1.161	.8617	.0934	.1084	10.703	9.222	4.377	40.365	10
11	1.178	.8489	.0843	.0993	11.863	10.071	4.851	48.855	11
12	1.196	.8364	.0767	.0917	13.041	10.907	5.322	58.054	12
13	1.214	.8240	.0702	.0852	14.237	11.731	5.791	67.943	13
14	1.232	.8118	.0647	.0797	15.450	12.543	6.258	78.496	14
15	1.250	.7999	.0599	.0749	16.682	13.343	6.722	89.694	15
16	1.269	.7880	.0558	.0708	17.932	14.131	7.184	101.514	16
17	1.288	.7764	.0521	.0671	19.201	14.908	7.643	113.937	17
18	1.307	.7649	.0488	.0638	20.489	15.673	8.100	126.940	18
19	1.327	.7536	.0459	.0609	21.797	16.426	8.554	140.505	19
20	1.347	.7425	.0432	.0582	23.124	17.169	9.005	154.611	20
21	1.367	.7315	.0409	.0559	24.470	17.900	9.455	169.241	21
22	1.388	.7207	.0387	.0537	25.837	18.621	9.902	184.375	22
23	1.408	.7100	.0367	.0517	27.225	19.331	10.346	199.996	23
24	1.430	.6995	.0349	.0499	28.633	20.030	10.788	216.085	24
25	1.451	.6892	.0333	.0483	30.063	20.720	11.227	232.626	25
26	1.473	.6790	.0317	.0467	31.514	21.399	11.664	249.601	26
27	1.495	.6690	.0303	.0453	32.987	22.068	12.099	266.995	27
28	1.517	.6591	.0290	.0440	34.481	22.727	12.531	284.790	28
29	1.540	.6494	.0278	.0428	35.999	23.376	12.961	302.972	29
30	1.563	.6398	.0266	.0416	37.539	24.016	13.388	321.525	30
36	1.709	.5851	.0212	.0362	47.276	27.661	15.901	439.823	36
40	1.814	.5513	.0184	.0334	54.268	29.916	17.528	524.349	40
48	2.043	.4894	.0144	.0294	69.565	34.042	20.666	703.537	48
50	2.105	.4750	.0136	.0286	73.682	35.000	21.428	749.955	50
52	2.169	.4611	.0128	.0278	77.925	35.929	22.179	796.868	52
60	2.443	.4093	.0104	.0254	96.214	39.380	25.093	988.157	60
70	2.835	.3527	.00817	.0232	122.363	43.155	28.529	1 231.15	70
72	2.921	.3423	.00781	.0228	128.076	43.845	29.189	1 279.78	72
80	3.291	.3039	.00655	.0215	152.710	46.407	31.742	1 473.06	80
84	3.493	.2863	.00602	.0210	166.172	47.579	32.967	1 568.50	84
90	3.819	.2619	.00532	.0203	187.929	49.210	34.740	1 709.53	90
96	4.176	.2395	.00472	.0197	211.719	50.702	36.438	1 847.46	96
100	4.432	.2256	.00437	.0194	228.802	51.625	37.529	1 937.43	100
104	4.704	.2126	.00405	.0190	246.932	52.494	38.589	2 025.69	104
120	5.969	.1675	.00302	.0180	331.286	55.498	42.518	2 359.69	120
240	35.632	.0281	.00043	.0154	2 308.8	64.796	59.737	3 870.68	240
360	212.700	.00470	.00007	.0151	14 113.3	66.353	64.966	4 310.71	360
480	1 269.7	.00079	.00001	.0150	84 577.8	66.614	66.288	4 415.74	480

1³/4%

Compound Interest Factors

1³/4%

	Single Payment		Uniform Payment Series				Arithmetic Gradient		
	Compound Amount Factor	Present Worth Factor	Sinking Fund Factor	Capital Recovery Factor	Compound Amount Factor	Present Worth Factor	Gradient Uniform Series	Gradient Present Worth	
	Find F Given P	Find P Given F	Find A Given F	Find A Given P	Find F Given A	Find P Given A	Find A Given G	Find P Given G	
n	F/P	P/F	A/F	A/P	F/A	P/A	A/G	P/G	n
1	1.018	.9828	1.0000	1.0175	1.000	0.983	0	0	1
2	1.035	.9659	.4957	.5132	2.018	1.949	0.496	0.966	2
3	1.053	.9493	.3276	.3451	3.053	2.898	0.989	2.865	3
4	1.072	.9330	.2435	.2610	4.106	3.831	1.478	5.664	4
5	1.091	.9169	.1931	.2106	5.178	4.748	1.965	9.332	5
6	1.110	.9011	.1595	.1770	6.269	5.649	2.450	13.837	6
7	1.129	.8856	.1355	.1530	7.378	6.535	2.931	19.152	7
8	1.149	.8704	.1175	.1350	8.508	7.405	3.409	25.245	8
9	1.169	.8554	.1036	.1211	9.656	8.261	3.885	32.088	9
10	1.189	.8407	.0924	.1099	10.825	9.101	4.357	39.655	10
11	1.210	.8263	.0832	.1007	12.015	9.928	4.827	47.918	11
12	1.231	.8121	.0756	.0931	13.225	10.740	5.294	56.851	12
13	1.253	.7981	.0692	.0867	14.457	11.538	5.758	66.428	13
14	1.275	.7844	.0637	.0812	15.710	12.322	6.219	76.625	14
15	1.297	.7709	.0589	.0764	16.985	13.093	6.677	87.417	15
16	1.320	.7576	.0547	.0722	18.282	13.851	7.132	98.782	16
17	1.343	.7446	.0510	.0685	19.602	14.595	7.584	110.695	17
18	1.367	.7318	.0477	.0652	20.945	15.327	8.034	123.136	18
19	1.390	.7192	.0448	.0623	22.311	16.046	8.481	136.081	19
20	1.415	.7068	.0422	.0597	23.702	16.753	8.924	149.511	20
21	1.440	.6947	.0398	.0573	25.116	17.448	9.365	163.405	21
22	1.465	.6827	.0377	.0552	26.556	18.130	9.804	177.742	22
23	1.490	.6710	.0357	.0532	28.021	18.801	10.239	192.503	23
24	1.516	.6594	.0339	.0514	29.511	19.461	10.671	207.671	24
25	1.543	.6481	.0322	.0497	31.028	20.109	11.101	223.225	25
26	1.570	.6369	.0307	.0482	32.571	20.746	11.528	239.149	26
27	1.597	.6260	.0293	.0468	34.141	21.372	11.952	255.425	27
28	1.625	.6152	.0280	.0455	35.738	21.987	12.373	272.036	28
29	1.654	.6046	.0268	.0443	37.363	22.592	12.791	288.967	29
30	1.683	.5942	.0256	.0431	39.017	23.186	13.206	306.200	30
36	1.867	.5355	.0202	.0377	49.566	26.543	15.640	415.130	36
40	2.002	.4996	.0175	.0350	57.234	28.594	17.207	492.017	40
48	2.300	.4349	.0135	.0310	74.263	32.294	20.209	652.612	48
50	2.381	.4200	.0127	.0302	78.903	33.141	20.932	693.708	50
52	2.465	.4057	.0119	.0294	83.706	33.960	21.644	735.039	52
60	2.832	.3531	.00955	.0271	104.676	36.964	24.389	901.503	60
70	3.368	.2969	.00739	.0249	135.331	40.178	27.586	1 108.34	70
72	3.487	.2868	.00704	.0245	142.127	40.757	28.195	1 149.12	72
80	4.006	.2496	.00582	.0233	171.795	42.880	30.533	1 309.25	80
84	4.294	.2329	.00531	.0228	188.246	43.836	31.644	1 387.16	84
90	4.765	.2098	.00465	.0221	215.166	45.152	33.241	1 500.88	90
96	5.288	.1891	.00408	.0216	245.039	46.337	34.756	1 610.48	96
100	5.668	.1764	.00375	.0212	266.753	47.062	35.721	1 681.09	100
104	6.075	.1646	.00345	.0209	290.028	47.737	36.652	1 749.68	104
120	8.019	.1247	.00249	.0200	401.099	50.017	40.047	2 003.03	120
240	64.308	.0156	.00028	.0178	3 617.6	56.254	53.352	3 001.27	240
360	515.702	.00194	.00003	.0175	29 411.5	57.032	56.443	3 219.08	360
480	4 135.5	.00024		.0175	236 259.0	57.129	57.027	3 257.88	480

2% Compound Interest Factors 2%

	Single Payment		Uniform Payment Series				Arithmetic Gradient		
	Compound Amount Factor	Present Worth Factor	Sinking Fund Factor	Capital Recovery Factor	Compound Amount Factor	Present Worth Factor	Gradient Uniform Series	Gradient Present Worth	
	Find F Given P	Find P Given F	Find A Given F	Find A Given P	Find F Given A	Find P Given A	Find A Given G	Find P Given G	
n	F/P	P/F	A/F	A/P	F/A	P/A	A/G	P/G	n
1	1.020	.9804	1.0000	1.0200	1.000	0.980	0	0	1
2	1.040	.9612	.4951	.5151	2.020	1.942	0.495	0.961	2
3	1.061	.9423	.3268	.3468	3.060	2.884	0.987	2.846	3
4	1.082	.9238	.2426	.2626	4.122	3.808	1.475	5.617	4
5	1.104	.9057	.1922	.2122	5.204	4.713	1.960	9.240	5
6	1.126	.8880	.1585	.1785	6.308	5.601	2.442	13.679	6
7	1.149	.8706	.1345	.1545	7.434	6.472	2.921	18.903	7
8	1.172	.8535	.1165	.1365	8.583	7.325	3.396	24.877	8
9	1.195	.8368	.1025	.1225	9.755	8.162	3.868	31.571	9
10	1.219	.8203	.0913	.1113	10.950	8.983	4.337	38.954	10
11	1.243	.8043	.0822	.1022	12.169	9.787	4.802	46.996	11
12	1.268	.7885	.0746	.0946	13.412	10.575	5.264	55.669	12
13	1.294	.7730	.0681	.0881	14.680	11.348	5.723	64.946	13
14	1.319	.7579	.0626	.0826	15.974	12.106	6.178	74.798	14
15	1.346	.7430	.0578	.0778	17.293	12.849	6.631	85.200	15
16	1.373	.7284	.0537	.0737	18.639	13.578	7.080	96.127	16
17	1.400	.7142	.0500	.0700	20.012	14.292	7.526	107.553	17
18	1.428	.7002	.0467	.0667	21.412	14.992	7.968	119.456	18
19	1.457	.6864	.0438	.0638	22.840	15.678	8.407	131.812	19
20	1.486	.6730	.0412	.0612	24.297	16.351	8.843	144.598	20
21	1.516	.6598	.0388	.0588	25.783	17.011	9.276	157.793	21
22	1.546	.6468	.0366	.0566	27.299	17.658	9.705	171.377	22
23	1.577	.6342	.0347	.0547	28.845	18.292	10.132	185.328	23
24	1.608	.6217	.0329	.0529	30.422	18.914	10.555	199.628	24
25	1.641	.6095	.0312	.0512	32.030	19.523	10.974	214.256	25
26	1.673	.5976	.0297	.0497	33.671	20.121	11.391	229.196	26
27	1.707	.5859	.0283	.0483	35.344	20.707	11.804	244.428	27
28	1.741	.5744	.0270	.0470	37.051	21.281	12.214	259.936	28
29	1.776	.5631	.0258	.0458	38.792	21.844	12.621	275.703	29
30	1.811	.5521	.0247	.0447	40.568	22.396	13.025	291.713	30
36	2.040	.4902	.0192	.0392	51.994	25.489	15.381	392.036	36
40	2.208	.4529	.0166	.0366	60.402	27.355	16.888	461.989	40
48	2.587	.3865	.0126	.0326	79.353	30.673	19.755	605.961	48
50	2.692	.3715	.0118	.0318	84.579	31.424	20.442	642.355	50
52	2.800	.3571	.0111	.0311	90.016	32.145	21.116	678.779	52
60	3.281	.3048	.00877	.0288	114.051	34.761	23.696	823.692	60
70	4.000	.2500	.00667	.0267	149.977	37.499	26.663	999.829	70
72	4.161	.2403	.00633	.0263	158.056	37.984	27.223	1 034.050	72
80	4.875	.2051	.00516	.0252	193.771	39.744	29.357	1 166.781	80
84	5.277	.1895	.00468	.0247	213.865	40.525	30.361	1 230.413	84
90	5.943	.1683	.00405	.0240	247.155	41.587	31.793	1 322.164	90
96	6.693	.1494	.00351	.0235	284.645	42.529	33.137	1 409.291	96
100	7.245	.1380	.00320	.0232	312.230	43.098	33.986	1 464.747	100
104	7.842	.1275	.00292	.0229	342.090	43.624	34.799	1 518.082	104
120	10.765	.0929	.00205	.0220	488.255	45.355	37.711	1 710.411	120
240	115.887	.00863	.00017	.0202	5 744.4	49.569	47.911	2 374.878	240
360	1 247.5	.00080	.00002	.0200	62 326.8	49.960	49.711	2 483.567	360
480	13 429.8	.00007		.0200	671 442.0	49.996	49.964	2 498.027	480

2¹/₂%

Compound Interest Factors

2¹/₂%

	Single Payment		Uniform Payment Series				Arithmetic Gradient		
	Compound Amount Factor	Present Worth Factor	Sinking Fund Factor	Capital Recovery Factor	Compound Amount Factor	Present Worth Factor	Gradient Uniform Series	Gradient Present Worth	
	Find F Given P	Find P Given F	Find A Given F	Find A Given P	Find F Given A	Find P Given A	Find A Given G	Find P Given G	
n	F/P	P/F	A/F	A/P	F/A	P/A	A/G	P/G	n
1	1.025	.9756	1.0000	1.0250	1.000	0.976	0	0	1
2	1.051	.9518	.4938	.5188	2.025	1.927	0.494	0.952	2
3	1.077	.9286	.3251	.3501	3.076	2.856	0.984	2.809	3
4	1.104	.9060	.2408	.2658	4.153	3.762	1.469	5.527	4
5	1.131	.8839	.1902	.2152	5.256	4.646	1.951	9.062	5
6	1.160	.8623	.1566	.1816	6.388	5.508	2.428	13.374	6
7	1.189	.8413	.1325	.1575	7.547	6.349	2.901	18.421	7
8	1.218	.8207	.1145	.1395	8.736	7.170	3.370	24.166	8
9	1.249	.8007	.1005	.1255	9.955	7.971	3.835	30.572	9
10	1.280	.7812	.0893	.1143	11.203	8.752	4.296	37.603	10
11	1.312	.7621	.0801	.1051	12.483	9.514	4.753	45.224	11
12	1.345	.7436	.0725	.0975	13.796	10.258	5.206	53.403	12
13	1.379	.7254	.0660	.0910	15.140	10.983	5.655	62.108	13
14	1.413	.7077	.0605	.0855	16.519	11.691	6.100	71.309	14
15	1.448	.6905	.0558	.0808	17.932	12.381	6.540	80.975	15
16	1.485	.6736	.0516	.0766	19.380	13.055	6.977	91.080	16
17	1.522	.6572	.0479	.0729	20.865	13.712	7.409	101.595	17
18	1.560	.6412	.0447	.0697	22.386	14.353	7.838	112.495	18
19	1.599	.6255	.0418	.0668	23.946	14.979	8.262	123.754	19
20	1.639	.6103	.0391	.0641	25.545	15.589	8.682	135.349	20
21	1.680	.5954	.0368	.0618	27.183	16.185	9.099	147.257	21
22	1.722	.5809	.0346	.0596	28.863	16.765	9.511	159.455	22
23	1.765	.5667	.0327	.0577	30.584	17.332	9.919	171.922	23
24	1.809	.5529	.0309	.0559	32.349	17.885	10.324	184.638	24
25	1.854	.5394	.0293	.0543	34.158	18.424	10.724	197.584	25
26	1.900	.5262	.0278	.0528	36.012	18.951	11.120	210.740	26
27	1.948	.5134	.0264	.0514	37.912	19.464	11.513	224.088	27
28	1.996	.5009	.0251	.0501	39.860	19.965	11.901	237.612	28
29	2.046	.4887	.0239	.0489	41.856	20.454	12.286	251.294	29
30	2.098	.4767	.0228	.0478	43.903	20.930	12.667	265.120	30
31	2.150	.4651	.0217	.0467	46.000	21.395	13.044	279.073	31
32	2.204	.4538	.0208	.0458	48.150	24.849	13.417	293.140	32
33	2.259	.4427	.0199	.0449	50.354	22.292	13.786	307.306	33
34	2.315	.4319	.0190	.0440	52.613	22.724	14.151	321.559	34
35	2.373	.4214	.0182	.0432	54.928	23.145	14.512	335.886	35
40	2.685	.3724	.0148	.0398	67.402	25.103	16.262	408.221	40
45	3.038	.3292	.0123	.0373	81.516	26.833	17.918	480.806	45
50	3.437	.2909	.0103	.0353	97.484	28.362	19.484	552.607	50
55	3.889	.2572	.00865	.0337	115.551	29.714	20.961	622.827	55
60	4.400	.2273	.00735	.0324	135.991	30.909	22.352	690.865	60
65	4.978	.2009	.00628	.0313	159.118	31.965	23.660	756.280	65
70	5.632	.1776	.00540	.0304	185.284	32.898	24.888	818.763	70
75	6.372	.1569	.00465	.0297	214.888	33.723	26.039	878.114	75
80	7.210	.1387	.00403	.0290	248.382	34.452	27.117	934.217	80
85	8.157	.1226	.00349	.0285	286.278	35.096	28.123	987.026	85
90	9.229	.1084	.00304	.0280	329.154	35.666	29.063	1 036.54	90
95	10.442	.0958	.00265	.0276	377.663	36.169	29.938	1 082.83	95
100	11.814	.0846	.00231	.0273	432.548	36.614	30.752	1 125.97	100

3%   Compound Interest Factors   3%

	Single Payment		Uniform Payment Series				Arithmetic Gradient		
	Compound Amount Factor	Present Worth Factor	Sinking Fund Factor	Capital Recovery Factor	Compound Amount Factor	Present Worth Factor	Gradient Uniform Series	Gradient Present Worth	
	Find F Given P	Find P Given F	Find A Given F	Find A Given P	Find F Given A	Find P Given A	Find A Given G	Find P Given G	
n	F/P	P/F	A/F	A/P	F/A	P/A	A/G	P/G	n
1	1.030	.9709	1.0000	1.0300	1.000	0.971	0	0	1
2	1.061	.9426	.4926	.5226	2.030	1.913	0.493	0.943	2
3	1.093	.9151	.3235	.3535	3.091	2.829	0.980	2.773	3
4	1.126	.8885	.2390	.2690	4.184	3.717	1.463	5.438	4
5	1.159	.8626	.1884	.2184	5.309	4.580	1.941	8.889	5
6	1.194	.8375	.1546	.1846	6.468	5.417	2.414	13.076	6
7	1.230	.8131	.1305	.1605	7.662	6.230	2.882	17.955	7
8	1.267	.7894	.1125	.1425	8.892	7.020	3.345	23.481	8
9	1.305	.7664	.0984	.1284	10.159	7.786	3.803	29.612	9
10	1.344	.7441	.0872	.1172	11.464	8.530	4.256	36.309	10
11	1.384	.7224	.0781	.1081	12.808	9.253	4.705	43.533	11
12	1.426	.7014	.0705	.1005	14.192	9.954	5.148	51.248	12
13	1.469	.6810	.0640	.0940	15.618	10.635	5.587	59.419	13
14	1.513	.6611	.0585	.0885	17.086	11.296	6.021	68.014	14
15	1.558	.6419	.0538	.0838	18.599	11.938	6.450	77.000	15
16	1.605	.6232	.0496	.0796	20.157	12.561	6.874	86.348	16
17	1.653	.6050	.0460	.0760	21.762	13.166	7.294	96.028	17
18	1.702	.5874	.0427	.0727	23.414	13.754	7.708	106.014	18
19	1.754	.5703	.0398	.0698	25.117	14.324	8.118	116.279	19
20	1.806	.5537	.0372	.0672	26.870	14.877	8.523	126.799	20
21	1.860	.5375	.0349	.0649	28.676	15.415	8.923	137.549	21
22	1.916	.5219	.0327	.0627	30.537	15.937	9.319	148.509	22
23	1.974	.5067	.0308	.0608	32.453	16.444	9.709	159.656	23
24	2.033	.4919	.0290	.0590	34.426	16.936	10.095	170.971	24
25	2.094	.4776	.0274	.0574	36.459	17.413	10.477	182.433	25
26	2.157	.4637	.0259	.0559	38.553	17.877	10.853	194.026	26
27	2.221	.4502	.0246	.0546	40.710	18.327	11.226	205.731	27
28	2.288	.4371	.0233	.0533	42.931	18.764	11.593	217.532	28
29	2.357	.4243	.0221	.0521	45.219	19.188	11.956	229.413	29
30	2.427	.4120	.0210	.0510	47.575	19.600	12.314	241.361	30
31	2.500	.4000	.0200	.0500	50.003	20.000	12.668	253.361	31
32	2.575	.3883	.0190	.0490	52.503	20.389	13.017	265.399	32
33	2.652	.3770	.0182	.0482	55.078	20.766	13.362	277.464	33
34	2.732	.3660	.0173	.0473	57.730	21.132	13.702	289.544	34
35	2.814	.3554	.0165	.0465	60.462	21.487	14.037	301.627	35
40	3.262	.3066	.0133	.0433	75.401	23.115	15.650	361.750	40
45	3.782	.2644	.0108	.0408	92.720	24.519	17.156	420.632	45
50	4.384	.2281	.00887	.0389	112.797	25.730	18.558	477.480	50
55	5.082	.1968	.00735	.0373	136.072	26.774	19.860	531.741	55
60	5.892	.1697	.00613	.0361	163.053	27.676	21.067	583.052	60
65	6.830	.1464	.00515	.0351	194.333	28.453	22.184	631.201	65
70	7.918	.1263	.00434	.0343	230.594	29.123	23.215	676.087	70
75	9.179	.1089	.00367	.0337	272.631	29.702	24.163	717.698	75
80	10.641	.0940	.00311	.0331	321.363	30.201	25.035	756.086	80
85	12.336	.0811	.00265	.0326	377.857	30.631	25.835	791.353	85
90	14.300	.0699	.00226	.0323	443.349	31.002	26.567	823.630	90
95	16.578	.0603	.00193	.0319	519.272	31.323	27.235	853.074	95
100	19.219	.0520	.00165	.0316	607.287	31.599	27.844	879.854	100

3¹/₂%

Compound Interest Factors

3¹/₂%

	Single Payment		Uniform Payment Series				Arithmetic Gradient		
	Compound Amount Factor	Present Worth Factor	Sinking Fund Factor	Capital Recovery Factor	Compound Amount Factor	Present Worth Factor	Gradient Uniform Series	Gradient Present Worth	
	Find F Given P	Find P Given F	Find A Given F	Find A Given P	Find F Given A	Find P Given A	Find A Given G	Find P Given G	
n	F/P	P/F	A/F	A/P	F/A	P/A	A/G	P/G	n
1	1.035	.9662	1.0000	1.0350	1.000	0.966	0	0	1
2	1.071	.9335	.4914	.5264	2.035	1.900	0.491	0.933	2
3	1.109	.9019	.3219	.3569	3.106	2.802	0.977	2.737	3
4	1.148	.8714	.2373	.2723	4.215	3.673	1.457	5.352	4
5	1.188	.8420	.1865	.2215	5.362	4.515	1.931	8.719	5
6	1.229	.8135	.1527	.1877	6.550	5.329	2.400	12.787	6
7	1.272	.7860	.1285	.1635	7.779	6.115	2.862	17.503	7
8	1.317	.7594	.1105	.1455	9.052	6.874	3.320	22.819	8
9	1.363	.7337	.0964	.1314	10.368	7.608	3.771	28.688	9
10	1.411	.7089	.0852	.1202	11.731	8.317	4.217	35.069	10
11	1.460	.6849	.0761	.1111	13.142	9.002	4.657	41.918	11
12	1.511	.6618	.0685	.1035	14.602	9.663	5.091	49.198	12
13	1.564	.6394	.0621	.0971	16.113	10.303	5.520	56.871	13
14	1.619	.6178	.0566	.0916	17.677	10.921	5.943	64.902	14
15	1.675	.5969	.0518	.0868	19.296	11.517	6.361	73.258	15
16	1.734	.5767	.0477	.0827	20.971	12.094	6.773	81.909	16
17	1.795	.5572	.0440	.0790	22.705	12.651	7.179	90.824	17
18	1.857	.5384	.0408	.0758	24.500	13.190	7.580	99.976	18
19	1.922	.5202	.0379	.0729	26.357	13.710	7.975	109.339	19
20	1.990	.5026	.0354	.0704	28.280	14.212	8.365	118.888	20
21	2.059	.4856	.0330	.0680	30.269	14.698	8.749	128.599	21
22	2.132	.4692	.0309	.0659	32.329	15.167	9.128	138.451	22
23	2.206	.4533	.0290	.0640	34.460	15.620	9.502	148.423	23
24	2.283	.4380	.0273	.0623	36.666	16.058	9.870	158.496	24
25	2.363	.4231	.0257	.0607	38.950	16.482	10.233	168.652	25
26	2.446	.4088	.0242	.0592	41.313	16.890	10.590	178.873	26
27	2.532	.3950	.0229	.0579	43.759	17.285	10.942	189.143	27
28	2.620	.3817	.0216	.0566	46.291	17.667	11.289	199.448	28
29	2.712	.3687	.0204	.0554	48.911	18.036	11.631	209.773	29
30	2.807	.3563	.0194	.0544	51.623	18.392	11.967	220.105	30
31	2.905	.3442	.0184	.0534	54.429	18.736	12.299	230.432	31
32	3.007	.3326	.0174	.0524	57.334	19.069	12.625	240.742	32
33	3.112	.3213	.0166	.0516	60.341	19.390	12.946	251.025	33
34	3.221	.3105	.0158	.0508	63.453	19.701	13.262	261.271	34
35	3.334	.3000	.0150	.0500	66.674	20.001	13.573	271.470	35
40	3.959	.2526	.0118	.0468	84.550	21.355	15.055	321.490	40
45	4.702	.2127	.00945	.0445	105.781	22.495	16.417	369.307	45
50	5.585	.1791	.00763	.0426	130.998	23.456	17.666	414.369	50
55	6.633	.1508	.00621	.0412	160.946	24.264	18.808	456.352	55
60	7.878	.1269	.00509	.0401	196.516	24.945	19.848	495.104	60
65	9.357	.1069	.00419	.0392	238.762	25.518	20.793	530.598	65
70	11.113	.0900	.00346	.0385	288.937	26.000	21.650	562.895	70
75	13.199	.0758	.00287	.0379	348.529	26.407	22.423	592.121	75
80	15.676	.0638	.00238	.0374	419.305	26.749	23.120	618.438	80
85	18.618	.0537	.00199	.0370	503.365	27.037	23.747	642.036	85
90	22.112	.0452	.00166	.0367	603.202	27.279	24.308	663.118	90
95	26.262	.0381	.00139	.0364	721.778	27.483	24.811	681.890	95
100	31.191	.0321	.00116	.0362	862.608	27.655	25.259	698.554	100

4%   Compound Interest Factors   4%

	Single Payment		Uniform Payment Series				Arithmetic Gradient		
	Compound Amount Factor	Present Worth Factor	Sinking Fund Factor	Capital Recovery Factor	Compound Amount Factor	Present Worth Factor	Gradient Uniform Series	Gradient Present Worth	
	Find F Given P	Find P Given F	Find A Given F	Find A Given P	Find F Given A	Find P Given A	Find A Given G	Find P Given G	
n	F/P	P/F	A/F	A/P	F/A	P/A	A/G	P/G	n
1	1.040	.9615	1.0000	1.0400	1.000	0.962	0	0	1
2	1.082	.9246	.4902	.5302	2.040	1.886	0.490	0.925	2
3	1.125	.8890	.3203	.3603	3.122	2.775	0.974	2.702	3
4	1.170	.8548	.2355	.2755	4.246	3.630	1.451	5.267	4
5	1.217	.8219	.1846	.2246	5.416	4.452	1.922	8.555	5
6	1.265	.7903	.1508	.1908	6.633	5.242	2.386	12.506	6
7	1.316	.7599	.1266	.1666	7.898	6.002	2.843	17.066	7
8	1.369	.7307	.1085	.1485	9.214	6.733	3.294	22.180	8
9	1.423	.7026	.0945	.1345	10.583	7.435	3.739	27.801	9
10	1.480	.6756	.0833	.1233	12.006	8.111	4.177	33.881	10
11	1.539	.6496	.0741	.1141	13.486	8.760	4.609	40.377	11
12	1.601	.6246	.0666	.1066	15.026	9.385	5.034	47.248	12
13	1.665	.6006	.0601	.1001	16.627	9.986	5.453	54.454	13
14	1.732	.5775	.0547	.0947	18.292	10.563	5.866	61.962	14
15	1.801	.5553	.0499	.0899	20.024	11.118	6.272	69.735	15
16	1.873	.5339	.0458	.0858	21.825	11.652	6.672	77.744	16
17	1.948	.5134	.0422	.0822	23.697	12.166	7.066	85.958	17
18	2.026	.4936	.0390	.0790	25.645	12.659	7.453	94.350	18
19	2.107	.4746	.0361	.0761	27.671	13.134	7.834	102.893	19
20	2.191	.4564	.0336	.0736	29.778	13.590	8.209	111.564	20
21	2.279	.4388	.0313	.0713	31.969	14.029	8.578	120.341	21
22	2.370	.4220	.0292	.0692	34.248	14.451	8.941	129.202	22
23	2.465	.4057	.0273	.0673	36.618	14.857	9.297	138.128	23
24	2.563	.3901	.0256	.0656	39.083	15.247	9.648	147.101	24
25	2.666	.3751	.0240	.0640	41.646	15.622	9.993	156.104	25
26	2.772	.3607	.0226	.0626	44.312	15.983	10.331	165.121	26
27	2.883	.3468	.0212	.0612	47.084	16.330	10.664	174.138	27
28	2.999	.3335	.0200	.0600	49.968	16.663	10.991	183.142	28
29	3.119	.3207	.0189	.0589	52.966	16.984	11.312	192.120	29
30	3.243	.3083	.0178	.0578	56.085	17.292	11.627	201.062	30
31	3.373	.2965	.0169	.0569	59.328	17.588	11.937	209.955	31
32	3.508	.2851	.0159	.0559	62.701	17.874	12.241	218.792	32
33	3.648	.2741	.0151	.0551	66.209	18.148	12.540	227.563	33
34	3.794	.2636	.0143	.0543	69.858	18.411	12.832	236.260	34
35	3.946	.2534	.0136	.0536	73.652	18.665	13.120	244.876	35
40	4.801	.2083	.0105	.0505	95.025	19.793	14.476	286.530	40
45	5.841	.1712	.00826	.0483	121.029	20.720	15.705	325.402	45
50	7.107	.1407	.00655	.0466	152.667	21.482	16.812	361.163	50
55	8.646	.1157	.00523	.0452	191.159	22.109	17.807	393.689	55
60	10.520	.0951	.00420	.0442	237.990	22.623	18.697	422.996	60
65	12.799	.0781	.00339	.0434	294.968	23.047	19.491	449.201	65
70	15.572	.0642	.00275	.0427	364.290	23.395	20.196	472.479	70
75	18.945	.0528	.00223	.0422	448.630	23.680	20.821	493.041	75
80	23.050	.0434	.00181	.0418	551.244	23.915	21.372	511.116	80
85	28.044	.0357	.00148	.0415	676.089	24.109	21.857	526.938	85
90	34.119	.0293	.00121	.0412	827.981	24.267	22.283	540.737	90
95	41.511	.0241	.00099	.0410	1 012.8	24.398	22.655	552.730	95
100	50.505	.0198	.00081	.0408	1 237.6	24.505	22.980	563.125	100

4¹/₂%

Compound Interest Factors

4¹/₂%

	Single Payment		Uniform Payment Series				Arithmetic Gradient		
	Compound Amount Factor	Present Worth Factor	Sinking Fund Factor	Capital Recovery Factor	Compound Amount Factor	Present Worth Factor	Gradient Uniform Series	Gradient Present Worth	
	Find F Given P	Find P Given F	Find A Given F	Find A Given P	Find F Given A	Find P Given A	Find A Given G	Find P Given G	
n	F/P	P/F	A/F	A/P	F/A	P/A	A/G	P/G	n
1	1.045	.9569	1.0000	1.0450	1.000	0.957	0	0	1
2	1.092	.9157	.4890	.5340	2.045	1.873	0.489	0.916	2
3	1.141	.8763	.3188	.3638	3.137	2.749	0.971	2.668	3
4	1.193	.8386	.2337	.2787	4.278	3.588	1.445	5.184	4
5	1.246	.8025	.1828	.2278	5.471	4.390	1.912	8.394	5
6	1.302	.7679	.1489	.1939	6.717	5.158	2.372	12.233	6
7	1.361	.7348	.1247	.1697	8.019	5.893	2.824	16.642	7
8	1.422	.7032	.1066	.1516	9.380	6.596	3.269	21.564	8
9	1.486	.6729	.0926	.1376	10.802	7.269	3.707	26.948	9
10	1.553	.6439	.0814	.1264	12.288	7.913	4.138	32.743	10
11	1.623	.6162	.0722	.1172	13.841	8.529	4.562	38.905	11
12	1.696	.5897	.0647	.1097	15.464	9.119	4.978	45.391	12
13	1.772	.5643	.0583	.1033	17.160	9.683	5.387	52.163	13
14	1.852	.5400	.0528	.0978	18.932	10.223	5.789	59.182	14
15	1.935	.5167	.0481	.0931	20.784	10.740	6.184	66.416	15
16	2.022	.4945	.0440	.0890	22.719	11.234	6.572	73.833	16
17	2.113	.4732	.0404	.0854	24.742	11.707	6.953	81.404	17
18	2.208	.4528	.0372	.0822	26.855	12.160	7.327	89.102	18
19	2.308	.4333	.0344	.0794	29.064	12.593	7.695	96.901	19
20	2.412	.4146	.0319	.0769	31.371	13.008	8.055	104.779	20
21	2.520	.3968	.0296	.0746	33.783	13.405	8.409	112.715	21
22	2.634	.3797	.0275	.0725	36.303	13.784	8.755	120.689	22
23	2.752	.3634	.0257	.0707	38.937	14.148	9.096	128.682	23
24	2.876	.3477	.0240	.0690	41.689	14.495	9.429	136.680	24
25	3.005	.3327	.0224	.0674	44.565	14.828	9.756	144.665	25
26	3.141	.3184	.0210	.0660	47.571	15.147	10.077	152.625	26
27	3.282	.3047	.0197	.0647	50.711	15.451	10.391	160.547	27
28	3.430	.2916	.0185	.0635	53.993	15.743	10.698	168.420	28
29	3.584	.2790	.0174	.0624	57.423	16.022	10.999	176.232	29
30	3.745	.2670	.0164	.0614	61.007	16.289	11.295	183.975	30
31	3.914	.2555	.0154	.0604	64.752	16.544	11.583	191.640	31
32	4.090	.2445	.0146	.0596	68.666	16.789	11.866	199.220	32
33	4.274	.2340	.0137	.0587	72.756	17.023	12.143	206.707	33
34	4.466	.2239	.0130	.0580	77.030	17.247	12.414	214.095	34
35	4.667	.2143	.0123	.0573	81.497	17.461	12.679	221.380	35
40	5.816	.1719	.00934	.0543	107.030	18.402	13.917	256.098	40
45	7.248	.1380	.00720	.0522	138.850	19.156	15.020	287.732	45
50	9.033	.1107	.00560	.0506	178.503	19.762	15.998	316.145	50
55	11.256	.0888	.00439	.0494	227.918	20.248	16.860	341.375	55
60	14.027	.0713	.00345	.0485	289.497	20.638	17.617	363.571	60
65	17.481	.0572	.00273	.0477	366.237	20.951	18.278	382.946	65
70	21.784	.0459	.00217	.0472	461.869	21.202	18.854	399.750	70
75	27.147	.0368	.00172	.0467	581.043	21.404	19.354	414.242	75
80	33.830	.0296	.00137	.0464	729.556	21.565	19.785	426.680	80
85	42.158	.0237	.00109	.0461	914.630	21.695	20.157	437.309	85
90	52.537	.0190	.00087	.0459	1 145.3	21.799	20.476	446.359	90
95	65.471	.0153	.00070	.0457	1 432.7	21.883	20.749	454.039	95
100	81.588	.0123	.00056	.0456	1 790.9	21.950	20.981	460.537	100

5% Compound Interest Factors 5%

	Single Payment		Uniform Payment Series				Arithmetic Gradient		
	Compound Amount Factor	Present Worth Factor	Sinking Fund Factor	Capital Recovery Factor	Compound Amount Factor	Present Worth Factor	Gradient Uniform Series	Gradient Present Worth	
	Find F Given P	Find P Given F	Find A Given F	Find A Given P	Find F Given A	Find P Given A	Find A Given G	Find P Given G	
n	F/P	P/F	A/F	A/P	F/A	P/A	A/G	P/G	n
1	1.050	.9524	1.0000	1.0500	1.000	0.952	0	0	1
2	1.102	.9070	.4878	.5378	2.050	1.859	0.488	0.907	2
3	1.158	.8638	.3172	.3672	3.152	2.723	0.967	2.635	3
4	1.216	.8227	.2320	.2820	4.310	3.546	1.439	5.103	4
5	1.276	.7835	.1810	.2310	5.526	4.329	1.902	8.237	5
6	1.340	.7462	.1470	.1970	6.802	5.076	2.358	11.968	6
7	1.407	.7107	.1228	.1728	8.142	5.786	2.805	16.232	7
8	1.477	.6768	.1047	.1547	9.549	6.463	3.244	20.970	8
9	1.551	.6446	.0907	.1407	11.027	7.108	3.676	26.127	9
10	1.629	.6139	.0795	.1295	12.578	7.722	4.099	31.652	10
11	1.710	.5847	.0704	.1204	14.207	8.306	4.514	37.499	11
12	1.796	.5568	.0628	.1128	15.917	8.863	4.922	43.624	12
13	1.886	.5303	.0565	.1065	17.713	9.394	5.321	49.988	13
14	1.980	.5051	.0510	.1010	19.599	9.899	5.713	56.553	14
15	2.079	.4810	.0463	.0963	21.579	10.380	6.097	63.288	15
16	2.183	.4581	.0423	.0923	23.657	10.838	6.474	70.159	16
17	2.292	.4363	.0387	.0887	25.840	11.274	6.842	77.140	17
18	2.407	.4155	.0355	.0855	28.132	11.690	7.203	84.204	18
19	2.527	.3957	.0327	.0827	30.539	12.085	7.557	91.327	19
20	2.653	.3769	.0302	.0802	33.066	12.462	7.903	98.488	20
21	2.786	.3589	.0280	.0780	35.719	12.821	8.242	105.667	21
22	2.925	.3419	.0260	.0760	38.505	13.163	8.573	112.846	22
23	3.072	.3256	.0241	.0741	41.430	13.489	8.897	120.008	23
24	3.225	.3101	.0225	.0725	44.502	13.799	9.214	127.140	24
25	3.386	.2953	.0210	.0710	47.727	14.094	9.524	134.227	25
26	3.556	.2812	.0196	.0696	51.113	14.375	9.827	141.258	26
27	3.733	.2678	.0183	.0683	54.669	14.643	10.122	148.222	27
28	3.920	.2551	.0171	.0671	58.402	14.898	10.411	155.110	28
29	4.116	.2429	.0160	.0660	62.323	15.141	10.694	161.912	29
30	4.322	.2314	.0151	.0651	66.439	15.372	10.969	168.622	30
31	4.538	.2204	.0141	.0641	70.761	15.593	11.238	175.233	31
32	4.765	.2099	.0133	.0633	75.299	15.803	11.501	181.739	32
33	5.003	.1999	.0125	.0625	80.063	16.003	11.757	188.135	33
34	5.253	.1904	.0118	.0618	85.067	16.193	12.006	194.416	34
35	5.516	.1813	.0111	.0611	90.320	16.374	12.250	200.580	35
40	7.040	.1420	.00828	.0583	120.799	17.159	13.377	229.545	40
45	8.985	.1113	.00626	.0563	159.699	17.774	14.364	255.314	45
50	11.467	.0872	.00478	.0548	209.347	18.256	15.223	277.914	50
55	14.636	.0683	.00367	.0537	272.711	18.633	15.966	297.510	55
60	18.679	.0535	.00283	.0528	353.582	18.929	16.606	314.343	60
65	23.840	.0419	.00219	.0522	456.795	19.161	17.154	328.691	65
70	30.426	.0329	.00170	.0517	588.525	19.343	17.621	340.841	70
75	38.832	.0258	.00132	.0513	756.649	19.485	18.018	351.072	75
80	49.561	.0202	.00103	.0510	971.222	19.596	18.353	359.646	80
85	63.254	.0158	.00080	.0508	1 245.1	19.684	18.635	366.800	85
90	80.730	.0124	.00063	.0506	1 594.6	19.752	18.871	372.749	90
95	103.034	.00971	.00049	.0505	2 040.7	19.806	19.069	377.677	95
100	131.500	.00760	.00038	.0504	2 610.0	19.848	19.234	381.749	100

6%   Compound Interest Factors   6%

	Single Payment		Uniform Payment Series				Arithmetic Gradient		
	Compound Amount Factor	Present Worth Factor	Sinking Fund Factor	Capital Recovery Factor	Compound Amount Factor	Present Worth Factor	Gradient Uniform Series	Gradient Present Worth	
	Find F Given P	Find P Given F	Find A Given F	Find A Given P	Find F Given A	Find P Given A	Find A Given G	Find P Given G	
n	F/P	P/F	A/F	A/P	F/A	P/A	A/G	P/G	n
1	1.060	.9434	1.0000	1.0600	1.000	0.943	0	0	1
2	1.124	.8900	.4854	.5454	2.060	1.833	0.485	0.890	2
3	1.191	.8396	.3141	.3741	3.184	2.673	0.961	2.569	3
4	1.262	.7921	.2286	.2886	4.375	3.465	1.427	4.945	4
5	1.338	.7473	.1774	.2374	5.637	4.212	1.884	7.934	5
6	1.419	.7050	.1434	.2034	6.975	4.917	2.330	11.459	6
7	1.504	.6651	.1191	.1791	8.394	5.582	2.768	15.450	7
8	1.594	.6274	.1010	.1610	9.897	6.210	3.195	19.841	8
9	1.689	.5919	.0870	.1470	11.491	6.802	3.613	24.577	9
10	1.791	.5584	.0759	.1359	13.181	7.360	4.022	29.602	10
11	1.898	.5268	.0668	.1268	14.972	7.887	4.421	34.870	11
12	2.012	.4970	.0593	.1193	16.870	8.384	4.811	40.337	12
13	2.133	.4688	.0530	.1130	18.882	8.853	5.192	45.963	13
14	2.261	.4423	.0476	.1076	21.015	9.295	5.564	51.713	14
15	2.397	.4173	.0430	.1030	23.276	9.712	5.926	57.554	15
16	2.540	.3936	.0390	.0990	25.672	10.106	6.279	63.459	16
17	2.693	.3714	.0354	.0954	28.213	10.477	6.624	69.401	17
18	2.854	.3503	.0324	.0924	30.906	10.828	6.960	75.357	18
19	3.026	.3305	.0296	.0896	33.760	11.158	7.287	81.306	19
20	3.207	.3118	.0272	.0872	36.786	11.470	7.605	87.230	20
21	3.400	.2942	.0250	.0850	39.993	11.764	7.915	93.113	21
22	3.604	.2775	.0230	.0830	43.392	12.042	8.217	98.941	22
23	3.820	.2618	.0213	.0813	46.996	12.303	8.510	104.700	23
24	4.049	.2470	.0197	.0797	50.815	12.550	8.795	110.381	24
25	4.292	.2330	.0182	.0782	54.864	12.783	9.072	115.973	25
26	4.549	.2198	.0169	.0769	59.156	13.003	9.341	121.468	26
27	4.822	.2074	.0157	.0757	63.706	13.211	9.603	126.860	27
28	5.112	.1956	.0146	.0746	68.528	13.406	9.857	132.142	28
29	5.418	.1846	.0136	.0736	73.640	13.591	10.103	137.309	29
30	5.743	.1741	.0126	.0726	79.058	13.765	10.342	142.359	30
31	6.088	.1643	.0118	.0718	84.801	13.929	10.574	147.286	31
32	6.453	.1550	.0110	.0710	90.890	14.084	10.799	152.090	32
33	6.841	.1462	.0103	.0703	97.343	14.230	11.017	156.768	33
34	7.251	.1379	.00960	.0696	104.184	14.368	11.228	161.319	34
35	7.686	.1301	.00897	.0690	111.435	14.498	11.432	165.743	35
40	10.286	.0972	.00646	.0665	154.762	15.046	12.359	185.957	40
45	13.765	.0727	.00470	.0647	212.743	15.456	13.141	203.109	45
50	18.420	.0543	.00344	.0634	290.335	15.762	13.796	217.457	50
55	24.650	.0406	.00254	.0625	394.171	15.991	14.341	229.322	55
60	32.988	.0303	.00188	.0619	533.126	16.161	14.791	239.043	60
65	44.145	.0227	.00139	.0614	719.080	16.289	15.160	246.945	65
70	59.076	.0169	.00103	.0610	967.928	16.385	15.461	253.327	70
75	79.057	.0126	.00077	.0608	1 300.9	16.456	15.706	258.453	75
80	105.796	.00945	.00057	.0606	1 746.6	16.509	15.903	262.549	80
85	141.578	.00706	.00043	.0604	2 343.0	16.549	16.062	265.810	85
90	189.464	.00528	.00032	.0603	3 141.1	16.579	16.189	268.395	90
95	253.545	.00394	.00024	.0602	4 209.1	16.601	16.290	270.437	95
100	339.300	.00295	.00018	.0602	5 638.3	16.618	16.371	272.047	100

7%

Compound Interest Factors

7%

	Single Payment		Uniform Payment Series				Arithmetic Gradient		
	Compound Amount Factor	Present Worth Factor	Sinking Fund Factor	Capital Recovery Factor	Compound Amount Factor	Present Worth Factor	Gradient Uniform Series	Gradient Present Worth	
	Find F Given P	Find P Given F	Find A Given F	Find A Given P	Find F Given A	Find P Given A	Find A Given G	Find P Given G	
n	F/P	P/F	A/F	A/P	F/A	P/A	A/G	P/G	n
1	1.070	.9346	1.0000	1.0700	1.000	0.935	0	0	1
2	1.145	.8734	.4831	.5531	2.070	1.808	0.483	0.873	2
3	1.225	.8163	.3111	.3811	3.215	2.624	0.955	2.506	3
4	1.311	.7629	.2252	.2952	4.440	3.387	1.416	4.795	4
5	1.403	.7130	.1739	.2439	5.751	4.100	1.865	7.647	5
6	1.501	.6663	.1398	.2098	7.153	4.767	2.303	10.978	6
7	1.606	.6227	.1156	.1856	8.654	5.389	2.730	14.715	7
8	1.718	.5820	.0975	.1675	10.260	5.971	3.147	18.789	8
9	1.838	.5439	.0835	.1535	11.978	6.515	3.552	23.140	9
10	1.967	.5083	.0724	.1424	13.816	7.024	3.946	27.716	10
11	2.105	.4751	.0634	.1334	15.784	7.499	4.330	32.467	11
12	2.252	.4440	.0559	.1259	17.888	7.943	4.703	37.351	12
13	2.410	.4150	.0497	.1197	20.141	8.358	5.065	42.330	13
14	2.579	.3878	.0443	.1143	22.551	8.745	5.417	47.372	14
15	2.759	.3624	.0398	.1098	25.129	9.108	5.758	52.446	15
16	2.952	.3387	.0359	.1059	27.888	9.447	6.090	57.527	16
17	3.159	.3166	.0324	.1024	30.840	9.763	6.411	62.592	17
18	3.380	.2959	.0294	.0994	33.999	10.059	6.722	67.622	18
19	3.617	.2765	.0268	.0968	37.379	10.336	7.024	72.599	19
20	3.870	.2584	.0244	.0944	40.996	10.594	7.316	77.509	20
21	4.141	.2415	.0223	.0923	44.865	10.836	7.599	82.339	21
22	4.430	.2257	.0204	.0904	49.006	11.061	7.872	87.079	22
23	4.741	.2109	.0187	.0887	53.436	11.272	8.137	91.720	23
24	5.072	.1971	.0172	.0872	58.177	11.469	8.392	96.255	24
25	5.427	.1842	.0158	.0858	63.249	11.654	8.639	100.677	25
26	5.807	.1722	.0146	.0846	68.677	11.826	8.877	104.981	26
27	6.214	.1609	.0134	.0834	74.484	11.987	9.107	109.166	27
28	6.649	.1504	.0124	.0824	80.698	12.137	9.329	113.227	28
29	7.114	.1406	.0114	.0814	87.347	12.278	9.543	117.162	29
30	7.612	.1314	.0106	.0806	94.461	12.409	9.749	120.972	30
31	8.145	.1228	.00980	.0798	102.073	12.532	9.947	124.655	31
32	8.715	.1147	.00907	.0791	110.218	12.647	10.138	128.212	32
33	9.325	.1072	.00841	.0784	118.934	12.754	10.322	131.644	33
34	9.978	.1002	.00780	.0778	128.259	12.854	10.499	134.951	34
35	10.677	.0937	.00723	.0772	138.237	12.948	10.669	138.135	35
40	14.974	.0668	.00501	.0750	199.636	13.332	11.423	152.293	40
45	21.002	.0476	.00350	.0735	285.750	13.606	12.036	163.756	45
50	29.457	.0339	.00246	.0725	406.530	13.801	12.529	172.905	50
55	41.315	.0242	.00174	.0717	575.930	13.940	12.921	180.124	55
60	57.947	.0173	.00123	.0712	813.523	14.039	13.232	185.768	60
65	81.273	.0123	.00087	.0709	1 146.8	14.110	13.476	190.145	65
70	113.990	.00877	.00062	.0706	1 614.1	14.160	13.666	193.519	70
75	159.877	.00625	.00044	.0704	2 269.7	14.196	13.814	196.104	75
80	224.235	.00446	.00031	.0703	3 189.1	14.222	13.927	198.075	80
85	314.502	.00318	.00022	.0702	4 478.6	14.240	14.015	199.572	85
90	441.105	.00227	.00016	.0702	6 287.2	14.253	14.081	200.704	90
95	618.673	.00162	.00011	.0701	8 823.9	14.263	14.132	201.558	95
100	867.720	.00115	.00008	.0701	12 381.7	14.269	14.170	202.200	100

8%

Compound Interest Factors

8%

	Single Payment		Uniform Payment Series				Arithmetic Gradient		
	Compound Amount Factor	Present Worth Factor	Sinking Fund Factor	Capital Recovery Factor	Compound Amount Factor	Present Worth Factor	Gradient Uniform Series	Gradient Present Worth	
	Find F Given P	Find P Given F	Find A Given F	Find A Given P	Find F Given A	Find P Given A	Find A Given G	Find P Given G	
n	F/P	P/F	A/F	A/P	F/A	P/A	A/G	P/G	n
1	1.080	.9259	1.0000	1.0800	1.000	0.926	0	0	1
2	1.166	.8573	.4808	.5608	2.080	1.783	0.481	0.857	2
3	1.260	.7938	.3080	.3880	3.246	2.577	0.949	2.445	3
4	1.360	.7350	.2219	.3019	4.506	3.312	1.404	4.650	4
5	1.469	.6806	.1705	.2505	5.867	3.993	1.846	7.372	5
6	1.587	.6302	.1363	.2163	7.336	4.623	2.276	10.523	6
7	1.714	.5835	.1121	.1921	8.923	5.206	2.694	14.024	7
8	1.851	.5403	.0940	.1740	10.637	5.747	3.099	17.806	8
9	1.999	.5002	.0801	.1601	12.488	6.247	3.491	21.808	9
10	2.159	.4632	.0690	.1490	14.487	6.710	3.871	25.977	10
11	2.332	.4289	.0601	.1401	16.645	7.139	4.240	30.266	11
12	2.518	.3971	.0527	.1327	18.977	7.536	4.596	34.634	12
13	2.720	.3677	.0465	.1265	21.495	7.904	4.940	39.046	13
14	2.937	.3405	.0413	.1213	24.215	8.244	5.273	43.472	14
15	3.172	.3152	.0368	.1168	27.152	8.559	5.594	47.886	15
16	3.426	.2919	.0330	.1130	30.324	8.851	5.905	52.264	16
17	3.700	.2703	.0296	.1096	33.750	9.122	6.204	56.588	17
18	3.996	.2502	.0267	.1067	37.450	9.372	6.492	60.843	18
19	4.316	.2317	.0241	.1041	41.446	9.604	6.770	65.013	19
20	4.661	.2145	.0219	.1019	45.762	9.818	7.037	69.090	20
21	5.034	.1987	.0198	.0998	50.423	10.017	7.294	73.063	21
22	5.437	.1839	.0180	.0980	55.457	10.201	7.541	76.926	22
23	5.871	.1703	.0164	.0964	60.893	10.371	7.779	80.673	23
24	6.341	.1577	.0150	.0950	66.765	10.529	8.007	84.300	24
25	6.848	.1460	.0137	.0937	73.106	10.675	8.225	87.804	25
26	7.396	.1352	.0125	.0925	79.954	10.810	8.435	91.184	26
27	7.988	.1252	.0114	.0914	87.351	10.935	8.636	94.439	27
28	8.627	.1159	.0105	.0905	95.339	11.051	8.829	97.569	28
29	9.317	.1073	.00962	.0896	103.966	11.158	9.013	100.574	29
30	10.063	.0994	.00883	.0888	113.283	11.258	9.190	103.456	30
31	10.868	.0920	.00811	.0881	123.346	11.350	9.358	106.216	31
32	11.737	.0852	.00745	.0875	134.214	11.435	9.520	108.858	32
33	12.676	.0789	.00685	.0869	145.951	11.514	9.674	111.382	33
34	13.690	.0730	.00630	.0863	158.627	11.587	9.821	113.792	34
35	14.785	.0676	.00580	.0858	172.317	11.655	9.961	116.092	35
40	21.725	.0460	.00386	.0839	259.057	11.925	10.570	126.042	40
45	31.920	.0313	.00259	.0826	386.506	12.108	11.045	133.733	45
50	46.902	.0213	.00174	.0817	573.771	12.233	11.411	139.593	50
55	68.914	.0145	.00118	.0812	848.925	12.319	11.690	144.006	55
60	101.257	.00988	.00080	.0808	1 253.2	12.377	11.902	147.300	60
65	148.780	.00672	.00054	.0805	1 847.3	12.416	12.060	149.739	65
70	218.607	.00457	.00037	.0804	2 720.1	12.443	12.178	151.533	70
75	321.205	.00311	.00025	.0802	4 002.6	12.461	12.266	152.845	75
80	471.956	.00212	.00017	.0802	5 887.0	12.474	12.330	153.800	80
85	693.458	.00144	.00012	.0801	8 655.7	12.482	12.377	154.492	85
90	1 018.9	.00098	.00008	.0801	12 724.0	12.488	12.412	154.993	90
95	1 497.1	.00067	.00005	.0801	18 701.6	12.492	12.437	155.352	95
100	2 199.8	.00045	.00004	.0800	27 484.6	12.494	12.455	155.611	100

9%   Compound Interest Factors   9%

| | Single Payment | | Uniform Payment Series | | | | Arithmetic Gradient | | |
| | Compound Amount Factor | Present Worth Factor | Sinking Fund Factor | Capital Recovery Factor | Compound Amount Factor | Present Worth Factor | Gradient Uniform Series | Gradient Present Worth | |
n	Find F Given P F/P	Find P Given F P/F	Find A Given F A/F	Find A Given P A/P	Find F Given A F/A	Find P Given A P/A	Find A Given G A/G	Find P Given G P/G	n
1	1.090	.9174	1.0000	1.0900	1.000	0.917	0	0	1
2	1.188	.8417	.4785	.5685	2.090	1.759	0.478	0.842	2
3	1.295	.7722	.3051	.3951	3.278	2.531	0.943	2.386	3
4	1.412	.7084	.2187	.3087	4.573	3.240	1.393	4.511	4
5	1.539	.6499	.1671	.2571	5.985	3.890	1.828	7.111	5
6	1.677	.5963	.1329	.2229	7.523	4.486	2.250	10.092	6
7	1.828	.5470	.1087	.1987	9.200	5.033	2.657	13.375	7
8	1.993	.5019	.0907	.1807	11.028	5.535	3.051	16.888	8
9	2.172	.4604	.0768	.1668	13.021	5.995	3.431	20.571	9
10	2.367	.4224	.0658	.1558	15.193	6.418	3.798	24.373	10
11	2.580	.3875	.0569	.1469	17.560	6.805	4.151	28.248	11
12	2.813	.3555	.0497	.1397	20.141	7.161	4.491	32.159	12
13	3.066	.3262	.0436	.1336	22.953	7.487	4.818	36.073	13
14	3.342	.2992	.0384	.1284	26.019	7.786	5.133	39.963	14
15	3.642	.2745	.0341	.1241	29.361	8.061	5.435	43.807	15
16	3.970	.2519	.0303	.1203	33.003	8.313	5.724	47.585	16
17	4.328	.2311	.0270	.1170	36.974	8.544	6.002	51.282	17
18	4.717	.2120	.0242	.1142	41.301	8.756	6.269	54.886	18
19	5.142	.1945	.0217	.1117	46.019	8.950	6.524	58.387	19
20	5.604	.1784	.0195	.1095	51.160	9.129	6.767	61.777	20
21	6.109	.1637	.0176	.1076	56.765	9.292	7.001	65.051	21
22	6.659	.1502	.0159	.1059	62.873	9.442	7.223	68.205	22
23	7.258	.1378	.0144	.1044	69.532	9.580	7.436	71.236	23
24	7.911	.1264	.0130	.1030	76.790	9.707	7.638	74.143	24
25	8.623	.1160	.0118	.1018	84.701	9.823	7.832	76.927	25
26	9.399	.1064	.0107	.1007	93.324	9.929	8.016	79.586	26
27	10.245	.0976	.00973	.0997	102.723	10.027	8.191	82.124	27
28	11.167	.0895	.00885	.0989	112.968	10.116	8.357	84.542	28
29	12.172	.0822	.00806	.0981	124.136	10.198	8.515	86.842	29
30	13.268	.0754	.00734	.0973	136.308	10.274	8.666	89.028	30
31	14.462	.0691	.00669	.0967	149.575	10.343	8.808	91.102	31
32	15.763	.0634	.00610	.0961	164.037	10.406	8.944	93.069	32
33	17.182	.0582	.00556	.0956	179.801	10.464	9.072	94.931	33
34	18.728	.0534	.00508	.0951	196.983	10.518	9.193	96.693	34
35	20.414	.0490	.00464	.0946	215.711	10.567	9.308	98.359	35
40	31.409	.0318	.00296	.0930	337.883	10.757	9.796	105.376	40
45	48.327	.0207	.00190	.0919	525.860	10.881	10.160	110.556	45
50	74.358	.0134	.00123	.0912	815.085	10.962	10.430	114.325	50
55	114.409	.00874	.00079	.0908	1 260.1	11.014	10.626	117.036	55
60	176.032	.00568	.00051	.0905	1 944.8	11.048	10.768	118.968	60
65	270.847	.00369	.00033	.0903	2 998.3	11.070	10.870	120.334	65
70	416.731	.00240	.00022	.0902	4 619.2	11.084	10.943	121.294	70
75	641.193	.00156	.00014	.0901	7 113.3	11.094	10.994	121.965	75
80	986.555	.00101	.00009	.0901	10 950.6	11.100	11.030	122.431	80
85	1 517.9	.00066	.00006	.0901	16 854.9	11.104	11.055	122.753	85
90	2 335.5	.00043	.00004	.0900	25 939.3	11.106	11.073	122.976	90
95	3 593.5	.00028	.00003	.0900	39 916.8	11.108	11.085	123.129	95
100	5 529.1	.00018	.00002	.0900	61 422.9	11.109	11.093	123.233	100

10%　　　　　　Compound Interest Factors　　　　　　10%

	Single Payment		Uniform Payment Series				Arithmetic Gradient		
	Compound Amount Factor	Present Worth Factor	Sinking Fund Factor	Capital Recovery Factor	Compound Amount Factor	Present Worth Factor	Gradient Uniform Series	Gradient Present Worth	
	Find F Given P	Find P Given F	Find A Given F	Find A Given P	Find F Given A	Find P Given A	Find A Given G	Find P Given G	
n	F/P	P/F	A/F	A/P	F/A	P/A	A/G	P/G	n
1	1.100	.9091	1.0000	1.1000	1.000	0.909	0	0	1
2	1.210	.8264	.4762	.5762	2.100	1.736	0.476	0.826	2
3	1.331	.7513	.3021	.4021	3.310	2.487	0.937	2.329	3
4	1.464	.6830	.2155	.3155	4.641	3.170	1.381	4.378	4
5	1.611	.6209	.1638	.2638	6.105	3.791	1.810	6.862	5
6	1.772	.5645	.1296	.2296	7.716	4.355	2.224	9.684	6
7	1.949	.5132	.1054	.2054	9.487	4.868	2.622	12.763	7
8	2.144	.4665	.0874	.1874	11.436	5.335	3.004	16.029	8
9	2.358	.4241	.0736	.1736	13.579	5.759	3.372	19.421	9
10	2.594	.3855	.0627	.1627	15.937	6.145	3.725	22.891	10
11	2.853	.3505	.0540	.1540	18.531	6.495	4.064	26.396	11
12	3.138	.3186	.0468	.1468	21.384	6.814	4.388	29.901	12
13	3.452	.2897	.0408	.1408	24.523	7.103	4.699	33.377	13
14	3.797	.2633	.0357	.1357	27.975	7.367	4.996	36.801	14
15	4.177	.2394	.0315	.1315	31.772	7.606	5.279	40.152	15
16	4.595	.2176	.0278	.1278	35.950	7.824	5.549	43.416	16
17	5.054	.1978	.0247	.1247	40.545	8.022	5.807	46.582	17
18	5.560	.1799	.0219	.1219	45.599	8.201	6.053	49.640	18
19	6.116	.1635	.0195	.1195	51.159	8.365	6.286	52.583	19
20	6.728	.1486	.0175	.1175	57.275	8.514	6.508	55.407	20
21	7.400	.1351	.0156	.1156	64.003	8.649	6.719	58.110	21
22	8.140	.1228	.0140	.1140	71.403	8.772	6.919	60.689	22
23	8.954	.1117	.0126	.1126	79.543	8.883	7.108	63.146	23
24	9.850	.1015	.0113	.1113	88.497	8.985	7.288	65.481	24
25	10.835	.0923	.0102	.1102	98.347	9.077	7.458	67.696	25
26	11.918	.0839	.00916	.1092	109.182	9.161	7.619	69.794	26
27	13.110	.0763	.00826	.1083	121.100	9.237	7.770	71.777	27
28	14.421	.0693	.00745	.1075	134.210	9.307	7.914	73.650	28
29	15.863	.0630	.00673	.1067	148.631	9.370	8.049	75.415	29
30	17.449	.0573	.00608	.1061	164.494	9.427	8.176	77.077	30
31	19.194	.0521	.00550	.1055	181.944	9.479	8.296	78.640	31
32	21.114	.0474	.00497	.1050	201.138	9.526	8.409	80.108	32
33	23.225	.0431	.00450	.1045	222.252	9.569	8.515	81.486	33
34	25.548	.0391	.00407	.1041	245.477	9.609	8.615	82.777	34
35	28.102	.0356	.00369	.1037	271.025	9.644	8.709	83.987	35
40	45.259	.0221	.00226	.1023	442.593	9.779	9.096	88.953	40
45	72.891	.0137	.00139	.1014	718.905	9.863	9.374	92.454	45
50	117.391	.00852	.00086	.1009	1 163.9	9.915	9.570	94.889	50
55	189.059	.00529	.00053	.1005	1 880.6	9.947	9.708	96.562	55
60	304.482	.00328	.00033	.1003	3 034.8	9.967	9.802	97.701	60
65	490.371	.00204	.00020	.1002	4 893.7	9.980	9.867	98.471	65
70	789.748	.00127	.00013	.1001	7 887.5	9.987	9.911	98.987	70
75	1 271.9	.00079	.00008	.1001	12 709.0	9.992	9.941	99.332	75
80	2 048.4	.00049	.00005	.1000	20 474.0	9.995	9.961	99.561	80
85	3 299.0	.00030	.00003	.1000	32 979.7	9.997	9.974	99.712	85
90	5 313.0	.00019	.00002	.1000	53 120.3	9.998	9.983	99.812	90
95	8 556.7	.00012	.00001	.1000	85 556.9	9.999	9.989	99.877	95
100	13 780.6	.00007	.00001	.1000	137 796.3	9.999	9.993	99.920	100

12%

Compound Interest Factors

12%

	Single Payment		Uniform Payment Series				Arithmetic Gradient		
	Compound Amount Factor	Present Worth Factor	Sinking Fund Factor	Capital Recovery Factor	Compound Amount Factor	Present Worth Factor	Gradient Uniform Series	Gradient Present Worth	
	Find F Given P F/P	Find P Given F P/F	Find A Given F A/F	Find A Given P A/P	Find F Given A F/A	Find P Given A P/A	Find A Given G A/G	Find P Given G P/G	
n									n
1	1.120	.8929	1.0000	1.1200	1.000	0.893	0	0	1
2	1.254	.7972	.4717	.5917	2.120	1.690	0.472	0.797	2
3	1.405	.7118	.2963	.4163	3.374	2.402	0.925	2.221	3
4	1.574	.6355	.2092	.3292	4.779	3.037	1.359	4.127	4
5	1.762	.5674	.1574	.2774	6.353	3.605	1.775	6.397	5
6	1.974	.5066	.1232	.2432	8.115	4.111	2.172	8.930	6
7	2.211	.4523	.0991	.2191	10.089	4.564	2.551	11.644	7
8	2.476	.4039	.0813	.2013	12.300	4.968	2.913	14.471	8
9	2.773	.3606	.0677	.1877	14.776	5.328	3.257	17.356	9
10	3.106	.3220	.0570	.1770	17.549	5.650	3.585	20.254	10
11	3.479	.2875	.0484	.1684	20.655	5.938	3.895	23.129	11
12	3.896	.2567	.0414	.1614	24.133	6.194	4.190	25.952	12
13	4.363	.2292	.0357	.1557	28.029	6.424	4.468	28.702	13
14	4.887	.2046	.0309	.1509	32.393	6.628	4.732	31.362	14
15	5.474	.1827	.0268	.1468	37.280	6.811	4.980	33.920	15
16	6.130	.1631	.0234	.1434	42.753	6.974	5.215	36.367	16
17	6.866	.1456	.0205	.1405	48.884	7.120	5.435	38.697	17
18	7.690	.1300	.0179	.1379	55.750	7.250	5.643	40.908	18
19	8.613	.1161	.0158	.1358	63.440	7.366	5.838	42.998	19
20	9.646	.1037	.0139	.1339	72.052	7.469	6.020	44.968	20
21	10.804	.0926	.0122	.1322	81.699	7.562	6.191	46.819	21
22	12.100	.0826	.0108	.1308	92.503	7.645	6.351	48.554	22
23	13.552	.0738	.00956	.1296	104.603	7.718	6.501	50.178	23
24	15.179	.0659	.00846	.1285	118.155	7.784	6.641	51.693	24
25	17.000	.0588	.00750	.1275	133.334	7.843	6.771	53.105	25
26	19.040	.0525	.00665	.1267	150.334	7.896	6.892	54.418	26
27	21.325	.0469	.00590	.1259	169.374	7.943	7.005	55.637	27
28	23.884	.0419	.00524	.1252	190.699	7.984	7.110	56.767	28
29	26.750	.0374	.00466	.1247	214.583	8.022	7.207	57.814	29
30	29.960	.0334	.00414	.1241	241.333	8.055	7.297	58.782	30
31	33.555	.0298	.00369	.1237	271.293	8.085	7.381	59.676	31
32	37.582	.0266	.00328	.1233	304.848	8.112	7.459	60.501	32
33	42.092	.0238	.00292	.1229	342.429	8.135	7.530	61.261	33
34	47.143	.0212	.00260	.1226	384.521	8.157	7.596	61.961	34
35	52.800	.0189	.00232	.1223	431.663	8.176	7.658	62.605	35
40	93.051	.0107	.00130	.1213	767.091	8.244	7.899	65.116	40
45	163.988	.00610	.00074	.1207	1 358.2	8.283	8.057	66.734	45
50	289.002	.00346	.00042	.1204	2 400.0	8.304	8.160	67.762	50
55	509.321	.00196	.00024	.1202	4 236.0	8.317	8.225	68.408	55
60	897.597	.00111	.00013	.1201	7 471.6	8.324	8.266	68.810	60
65	1 581.9	.00063	.00008	.1201	13 173.9	8.328	8.292	69.058	65
70	2 787.8	.00036	.00004	.1200	23 223.3	8.330	8.308	69.210	70
75	4 913.1	.00020	.00002	.1200	40 933.8	8.332	8.318	69.303	75
80	8 658.5	.00012	.00001	.1200	72 145.7	8.332	8.324	69.359	80
85	15 259.2	.00007	.00001	.1200	127 151.7	8.333	8.328	69.393	85
90	26 891.9	.00004		.1200	224 091.1	8.333	8.330	69.414	90
95	47 392.8	.00002		.1200	394 931.4	8.333	8.331	69.426	95
100	83 522.3	.00001		.1200	696 010.5	8.333	8.332	69.434	100

15%   **Compound Interest Factors**   # 15%

	Single Payment		Uniform Payment Series				Arithmetic Gradient		
	Compound Amount Factor	Present Worth Factor	Sinking Fund Factor	Capital Recovery Factor	Compound Amount Factor	Present Worth Factor	Gradient Uniform Series	Gradient Present Worth	
	Find F Given P	Find P Given F	Find A Given F	Find A Given P	Find F Given A	Find P Given A	Find A Given G	Find P Given G	
n	F/P	P/F	A/F	A/P	F/A	P/A	A/G	P/G	n
1	1.150	.8696	1.0000	1.1500	1.000	0.870	0	0	1
2	1.322	.7561	.4651	.6151	2.150	1.626	0.465	0.756	2
3	1.521	.6575	.2880	.4380	3.472	2.283	0.907	2.071	3
4	1.749	.5718	.2003	.3503	4.993	2.855	1.326	3.786	4
5	2.011	.4972	.1483	.2983	6.742	3.352	1.723	5.775	5
6	2.313	.4323	.1142	.2642	8.754	3.784	2.097	7.937	6
7	2.660	.3759	.0904	.2404	11.067	4.160	2.450	10.192	7
8	3.059	.3269	.0729	.2229	13.727	4.487	2.781	12.481	8
9	3.518	.2843	.0596	.2096	16.786	4.772	3.092	14.755	9
10	4.046	.2472	.0493	.1993	20.304	5.019	3.383	16.979	10
11	4.652	.2149	.0411	.1911	24.349	5.234	3.655	19.129	11
12	5.350	.1869	.0345	.1845	29.002	5.421	3.908	21.185	12
13	6.153	.1625	.0291	.1791	34.352	5.583	4.144	23.135	13
14	7.076	.1413	.0247	.1747	40.505	5.724	4.362	24.972	14
15	8.137	.1229	.0210	.1710	47.580	5.847	4.565	26.693	15
16	9.358	.1069	.0179	.1679	55.717	5.954	4.752	28.296	16
17	10.761	.0929	.0154	.1654	65.075	6.047	4.925	29.783	17
18	12.375	.0808	.0132	.1632	75.836	6.128	5.084	31.156	18
19	14.232	.0703	.0113	.1613	88.212	6.198	5.231	32.421	19
20	16.367	.0611	.00976	.1598	102.444	6.259	5.365	33.582	20
21	18.822	.0531	.00842	.1584	118.810	6.312	5.488	34.645	21
22	21.645	.0462	.00727	.1573	137.632	6.359	5.601	35.615	22
23	24.891	.0402	.00628	.1563	159.276	6.399	5.704	36.499	23
24	28.625	.0349	.00543	.1554	184.168	6.434	5.798	37.302	24
25	32.919	.0304	.00470	.1547	212.793	6.464	5.883	38.031	25
26	37.857	.0264	.00407	.1541	245.712	6.491	5.961	38.692	26
27	43.535	.0230	.00353	.1535	283.569	6.514	6.032	39.289	27
28	50.066	.0200	.00306	.1531	327.104	6.534	6.096	39.828	28
29	57.575	.0174	.00265	.1527	377.170	6.551	6.154	40.315	29
30	66.212	.0151	.00230	.1523	434.745	6.566	6.207	40.753	30
31	76.144	.0131	.00200	.1520	500.957	6.579	6.254	41.147	31
32	87.565	.0114	.00173	.1517	577.100	6.591	6.297	41.501	32
33	100.700	.00993	.00150	.1515	664.666	6.600	6.336	41.818	33
34	115.805	.00864	.00131	.1513	765.365	6.609	6.371	42.103	34
35	133.176	.00751	.00113	.1511	881.170	6.617	6.402	42.359	35
40	267.864	.00373	.00056	.1506	1 779.1	6.642	6.517	43.283	40
45	538.769	.00186	.00028	.1503	3 585.1	6.654	6.583	43.805	45
50	1 083.7	.00092	.00014	.1501	7 217.7	6.661	6.620	44.096	50
55	2 179.6	.00046	.00007	.1501	14 524.1	6.664	6.641	44.256	55
60	4 384.0	.00023	.00003	.1500	29 220.0	6.665	6.653	44.343	60
65	8 817.8	.00011	.00002	.1500	58 778.6	6.666	6.659	44.390	65
70	17 735.7	.00006	.00001	.1500	118 231.5	6.666	6.663	44.416	70
75	35 672.9	.00003		.1500	237 812.5	6.666	6.665	44.429	75
80	71 750.9	.00001		.1500	478 332.6	6.667	6.666	44.436	80
85	144 316.7	.00001		.1500	962 104.4	6.667	6.666	44.440	85

18% Compound Interest Factors 18%

	Single Payment		Uniform Payment Series				Arithmetic Gradient		
	Compound Amount Factor	Present Worth Factor	Sinking Fund Factor	Capital Recovery Factor	Compound Amount Factor	Present Worth Factor	Gradient Uniform Series	Gradient Present Worth	
	Find F Given P	Find P Given F	Find A Given F	Find A Given P	Find F Given A	Find P Given A	Find A Given G	Find P Given G	
n	F/P	P/F	A/F	A/P	F/A	P/A	A/G	P/G	n
1	1.180	.8475	1.0000	1.1800	1.000	0.847	0	0	1
2	1.392	.7182	.4587	.6387	2.180	1.566	0.459	0.718	2
3	1.643	.6086	.2799	.4599	3.572	2.174	0.890	1.935	3
4	1.939	.5158	.1917	.3717	5.215	2.690	1.295	3.483	4
5	2.288	.4371	.1398	.3198	7.154	3.127	1.673	5.231	5
6	2.700	.3704	.1059	.2859	9.442	3.498	2.025	7.083	6
7	3.185	.3139	.0824	.2624	12.142	3.812	2.353	8.967	7
8	3.759	.2660	.0652	.2452	15.327	4.078	2.656	10.829	8
9	4.435	.2255	.0524	.2324	19.086	4.303	2.936	12.633	9
10	5.234	.1911	.0425	.2225	23.521	4.494	3.194	14.352	10
11	6.176	.1619	.0348	.2148	28.755	4.656	3.430	15.972	11
12	7.288	.1372	.0286	.2086	34.931	4.793	3.647	17.481	12
13	8.599	.1163	.0237	.2037	42.219	4.910	3.845	18.877	13
14	10.147	.0985	.0197	.1997	50.818	5.008	4.025	20.158	14
15	11.974	.0835	.0164	.1964	60.965	5.092	4.189	21.327	15
16	14.129	.0708	.0137	.1937	72.939	5.162	4.337	22.389	16
17	16.672	.0600	.0115	.1915	87.068	5.222	4.471	23.348	17
18	19.673	.0508	.00964	.1896	103.740	5.273	4.592	24.212	18
19	23.214	.0431	.00810	.1881	123.413	5.316	4.700	24.988	19
20	27.393	.0365	.00682	.1868	146.628	5.353	4.798	25.681	20
21	32.324	.0309	.00575	.1857	174.021	5.384	4.885	26.300	21
22	38.142	.0262	.00485	.1848	206.345	5.410	4.963	26.851	22
23	45.008	.0222	.00409	.1841	244.487	5.432	5.033	27.339	23
24	53.109	.0188	.00345	.1835	289.494	5.451	5.095	27.772	24
25	62.669	.0160	.00292	.1829	342.603	5.467	5.150	28.155	25
26	73.949	.0135	.00247	.1825	405.272	5.480	5.199	28.494	26
27	87.260	.0115	.00209	.1821	479.221	5.492	5.243	28.791	27
28	102.966	.00971	.00177	.1818	566.480	5.502	5.281	29.054	28
29	121.500	.00823	.00149	.1815	669.447	5.510	5.315	29.284	29
30	143.370	.00697	.00126	.1813	790.947	5.517	5.345	29.486	30
31	169.177	.00591	.00107	.1811	934.317	5.523	5.371	29.664	31
32	199.629	.00501	.00091	.1809	1 103.5	5.528	5.394	29.819	32
33	235.562	.00425	.00077	.1808	1 303.1	5.532	5.415	29.955	33
34	277.963	.00360	.00065	.1806	1 538.7	5.536	5.433	30.074	34
35	327.997	.00305	.00055	.1806	1 816.6	5.539	5.449	30.177	35
40	750.377	.00133	.00024	.1802	4 163.2	5.548	5.502	30.527	40
45	1 716.7	.00058	.00010	.1801	9 531.6	5.552	5.529	30.701	45
50	3 927.3	.00025	.00005	.1800	21 813.0	5.554	5.543	30.786	50
55	8 984.8	.00011	.00002	.1800	49 910.1	5.555	5.549	30.827	55
60	20 555.1	.00005	.00001	.1800	114 189.4	5.555	5.553	30.846	60
65	47 025.1	.00002		.1800	261 244.7	5.555	5.554	30.856	65
70	107 581.9	.00001		.1800	597 671.7	5.556	5.555	30.860	70

20% Compound Interest Factors 20%

	Single Payment		Uniform Payment Series				Arithmetic Gradient		
	Compound Amount Factor	Present Worth Factor	Sinking Fund Factor	Capital Recovery Factor	Compound Amount Factor	Present Worth Factor	Gradient Uniform Series	Gradient Present Worth	
	Find F Given P	Find P Given F	Find A Given F	Find A Given P	Find F Given A	Find P Given A	Find A Given G	Find P Given G	
n	F/P	P/F	A/F	A/P	F/A	P/A	A/G	P/G	n
1	1.200	.8333	1.0000	1.2000	1.000	0.833	0	0	1
2	1.440	.6944	.4545	.6545	2.200	1.528	0.455	0.694	2
3	1.728	.5787	.2747	.4747	3.640	2.106	0.879	1.852	3
4	2.074	.4823	.1863	.3863	5.368	2.589	1.274	3.299	4
5	2.488	.4019	.1344	.3344	7.442	2.991	1.641	4.906	5
6	2.986	.3349	.1007	.3007	9.930	3.326	1.979	6.581	6
7	3.583	.2791	.0774	.2774	12.916	3.605	2.290	8.255	7
8	4.300	.2326	.0606	.2606	16.499	3.837	2.576	9.883	8
9	5.160	.1938	.0481	.2481	20.799	4.031	2.836	11.434	9
10	6.192	.1615	.0385	.2385	25.959	4.192	3.074	12.887	10
11	7.430	.1346	.0311	.2311	32.150	4.327	3.289	14.233	11
12	8.916	.1122	.0253	.2253	39.581	4.439	3.484	15.467	12
13	10.699	.0935	.0206	.2206	48.497	4.533	3.660	16.588	13
14	12.839	.0779	.0169	.2169	59.196	4.611	3.817	17.601	14
15	15.407	.0649	.0139	.2139	72.035	4.675	3.959	18.509	15
16	18.488	.0541	.0114	.2114	87.442	4.730	4.085	19.321	16
17	22.186	.0451	.00944	.2094	105.931	4.775	4.198	20.042	17
18	26.623	.0376	.00781	.2078	128.117	4.812	4.298	20.680	18
19	31.948	.0313	.00646	.2065	154.740	4.843	4.386	21.244	19
20	38.338	.0261	.00536	.2054	186.688	4.870	4.464	21.739	20
21	46.005	.0217	.00444	.2044	225.026	4.891	4.533	22.174	21
22	55.206	.0181	.00369	.2037	271.031	4.909	4.594	22.555	22
23	66.247	.0151	.00307	.2031	326.237	4.925	4.647	22.887	23
24	79.497	.0126	.00255	.2025	392.484	4.937	4.694	23.176	24
25	95.396	.0105	.00212	.2021	471.981	4.948	4.735	23.428	25
26	114.475	.00874	.00176	.2018	567.377	4.956	4.771	23.646	26
27	137.371	.00728	.00147	.2015	681.853	4.964	4.802	23.835	27
28	164.845	.00607	.00122	.2012	819.223	4.970	4.829	23.999	28
29	197.814	.00506	.00102	.2010	984.068	4.975	4.853	24.141	29
30	237.376	.00421	.00085	.2008	1 181.9	4.979	4.873	24.263	30
31	284.852	.00351	.00070	.2007	1 419.3	4.982	4.891	24.368	31
32	341.822	.00293	.00059	.2006	1 704.1	4.985	4.906	24.459	32
33	410.186	.00244	.00049	.2005	2 045.9	4.988	4.919	24.537	33
34	492.224	.00203	.00041	.2004	2 456.1	4.990	4.931	24.604	34
35	590.668	.00169	.00034	.2003	2 948.3	4.992	4.941	24.661	35
40	1 469.8	.00068	.00014	.2001	7 343.9	4.997	4.973	24.847	40
45	3 657.3	.00027	.00005	.2001	18 281.3	4.999	4.988	24.932	45
50	9 100.4	.00011	.00002	.2000	45 497.2	4.999	4.995	24.970	50
55	22 644.8	.00004	.00001	.2000	113 219.0	5.000	4.998	24.987	55
60	56 347.5	.00002		.2000	281 732.6	5.000	4.999	24.994	60

25%   Compound Interest Factors   25%

	Single Payment		Uniform Payment Series				Arithmetic Gradient		
	Compound Amount Factor	Present Worth Factor	Sinking Fund Factor	Capital Recovery Factor	Compound Amount Factor	Present Worth Factor	Gradient Uniform Series	Gradient Present Worth	
	Find F Given P	Find P Given F	Find A Given F	Find A Given P	Find F Given A	Find P Given A	Find A Given G	Find P Given G	
n	F/P	P/F	A/F	A/P	F/A	P/A	A/G	P/G	n
1	1.250	.8000	1.0000	1.2500	1.000	0.800	0	0	1
2	1.563	.6400	.4444	.6944	2.250	1.440	0.444	0.640	2
3	1.953	.5120	.2623	.5123	3.813	1.952	0.852	1.664	3
4	2.441	.4096	.1734	.4234	5.766	2.362	1.225	2.893	4
5	3.052	.3277	.1218	.3718	8.207	2.689	1.563	4.204	5
6	3.815	.2621	.0888	.3388	11.259	2.951	1.868	5.514	6
7	4.768	.2097	.0663	.3163	15.073	3.161	2.142	6.773	7
8	5.960	.1678	.0504	.3004	19.842	3.329	2.387	7.947	8
9	7.451	.1342	.0388	.2888	25.802	3.463	2.605	9.021	9
10	9.313	.1074	.0301	.2801	33.253	3.571	2.797	9.987	10
11	11.642	.0859	.0235	.2735	42.566	3.656	2.966	10.846	11
12	14.552	.0687	.0184	.2684	54.208	3.725	3.115	11.602	12
13	18.190	.0550	.0145	.2645	68.760	3.780	3.244	12.262	13
14	22.737	.0440	.0115	.2615	86.949	3.824	3.356	12.833	14
15	28.422	.0352	.00912	.2591	109.687	3.859	3.453	13.326	15
16	35.527	.0281	.00724	.2572	138.109	3.887	3.537	13.748	16
17	44.409	.0225	.00576	.2558	173.636	3.910	3.608	14.108	17
18	55.511	.0180	.00459	.2546	218.045	3.928	3.670	14.415	18
19	69.389	.0144	.00366	.2537	273.556	3.942	3.722	14.674	19
20	86.736	.0115	.00292	.2529	342.945	3.954	3.767	14.893	20
21	108.420	.00922	.00233	.2523	429.681	3.963	3.805	15.078	21
22	135.525	.00738	.00186	.2519	538.101	3.970	3.836	15.233	22
23	169.407	.00590	.00148	.2515	673.626	3.976	3.863	15.362	23
24	211.758	.00472	.00119	.2512	843.033	3.981	3.886	15.471	24
25	264.698	.00378	.00095	.2509	1 054.8	3.985	3.905	15.562	25
26	330.872	.00302	.00076	.2508	1 319.5	3.988	3.921	15.637	26
27	413.590	.00242	.00061	.2506	1 650.4	3.990	3.935	15.700	27
28	516.988	.00193	.00048	.2505	2 064.0	3.992	3.946	15.752	28
29	646.235	.00155	.00039	.2504	2 580.9	3.994	3.955	15.796	29
30	807.794	.00124	.00031	.2503	3 227.2	3.995	3.963	15.832	30
31	1 009.7	.00099	.00025	.2502	4 035.0	3.996	3.969	15.861	31
32	1 262.2	.00079	.00020	.2502	5 044.7	3.997	3.975	15.886	32
33	1 577.7	.00063	.00016	.2502	6 306.9	3.997	3.979	15.906	33
34	1 972.2	.00051	.00013	.2501	7 884.6	3.998	3.983	15.923	34
35	2 465.2	.00041	.00010	.2501	9 856.8	3.998	3.986	15.937	35
40	7 523.2	.00013	.00003	.2500	30 088.7	3.999	3.995	15.977	40
45	22 958.9	.00004	.00001	.2500	91 831.5	4.000	3.998	15.991	45
50	70 064.9	.00001		.2500	280 255.7	4.000	3.999	15.997	50
55	213 821.2			.2500	855 280.7	4.000	4.000	15.999	55

30% Compound Interest Factors 30%

	Single Payment		Uniform Payment Series				Arithmetic Gradient		
	Compound Amount Factor	Present Worth Factor	Sinking Fund Factor	Capital Recovery Factor	Compound Amount Factor	Present Worth Factor	Gradient Uniform Series	Gradient Present Worth	
	Find F Given P	Find P Given F	Find A Given F	Find A Given P	Find F Given A	Find P Given A	Find A Given G	Find P Given G	
n	F/P	P/F	A/F	A/P	F/A	P/A	A/G	P/G	n
1	1.300	.7692	1.0000	1.3000	1.000	0.769	0	0	1
2	1.690	.5917	.4348	.7348	2.300	1.361	0.435	0.592	2
3	2.197	.4552	.2506	.5506	3.990	1.816	0.827	1.502	3
4	2.856	.3501	.1616	.4616	6.187	2.166	1.178	2.552	4
5	3.713	.2693	.1106	.4106	9.043	2.436	1.490	3.630	5
6	4.827	.2072	.0784	.3784	12.756	2.643	1.765	4.666	6
7	6.275	.1594	.0569	.3569	17.583	2.802	2.006	5.622	7
8	8.157	.1226	.0419	.3419	23.858	2.925	2.216	6.480	8
9	10.604	.0943	.0312	.3312	32.015	3.019	2.396	7.234	9
10	13.786	.0725	.0235	.3235	42.619	3.092	2.551	7.887	10
11	17.922	.0558	.0177	.3177	56.405	3.147	2.683	8.445	11
12	23.298	.0429	.0135	.3135	74.327	3.190	2.795	8.917	12
13	30.287	.0330	.0102	.3102	97.625	3.223	2.889	9.314	13
14	39.374	.0254	.00782	.3078	127.912	3.249	2.969	9.644	14
15	51.186	.0195	.00598	.3060	167.286	3.268	3.034	9.917	15
16	66.542	.0150	.00458	.3046	218.472	3.283	3.089	10.143	16
17	86.504	.0116	.00351	.3035	285.014	3.295	3.135	10.328	17
18	112.455	.00889	.00269	.3027	371.518	3.304	3.172	10.479	18
19	146.192	.00684	.00207	.3021	483.973	3.311	3.202	10.602	19
20	190.049	.00526	.00159	.3016	630.165	3.316	3.228	10.702	20
21	247.064	.00405	.00122	.3012	820.214	3.320	3.248	10.783	21
22	321.184	.00311	.00094	.3009	1 067.3	3.323	3.265	10.848	22
23	417.539	.00239	.00072	.3007	1 388.5	3.325	3.278	10.901	23
24	542.800	.00184	.00055	.3006	1 806.0	3.327	3.289	10.943	24
25	705.640	.00142	.00043	.3004	2 348.8	3.329	3.298	10.977	25
26	917.332	.00109	.00033	.3003	3 054.4	3.330	3.305	11.005	26
27	1 192.5	.00084	.00025	.3003	3 971.8	3.331	3.311	11.026	27
28	1 550.3	.00065	.00019	.3002	5 164.3	3.331	3.315	11.044	28
29	2 015.4	.00050	.00015	.3001	6 714.6	3.332	3.319	11.058	29
30	2 620.0	.00038	.00011	.3001	8 730.0	3.332	3.322	11.069	30
31	3 406.0	.00029	.00009	.3001	11 350.0	3.332	3.324	11.078	31
32	4 427.8	.00023	.00007	.3001	14 756.0	3.333	3.326	11.085	32
33	5 756.1	.00017	.00005	.3001	19 183.7	3.333	3.328	11.090	33
34	7 483.0	.00013	.00004	.3000	24 939.9	3.333	3.329	11.094	34
35	9 727.8	.00010	.00003	.3000	32 422.8	3.333	3.330	11.098	35
40	36 118.8	.00003	.00001	.3000	120 392.6	3.333	3.332	11.107	40
45	134 106.5	.00001		.3000	447 018.3	3.333	3.333	11.110	45

35%

Compound Interest Factors

35%

	Single Payment		Uniform Payment Series				Arithmetic Gradient		
	Compound Amount Factor	Present Worth Factor	Sinking Fund Factor	Capital Recovery Factor	Compound Amount Factor	Present Worth Factor	Gradient Uniform Series	Gradient Present Worth	
	Find F Given P	Find P Given F	Find A Given F	Find A Given P	Find F Given A	Find P Given A	Find A Given G	Find P Given G	
n	F/P	P/F	A/F	A/P	F/A	P/A	A/G	P/G	n
1	1.350	.7407	1.0000	1.3500	1.000	0.741	0	0	1
2	1.822	.5487	.4255	.7755	2.350	1.289	0.426	0.549	2
3	2.460	.4064	.2397	.5897	4.173	1.696	0.803	1.362	3
4	3.322	.3011	.1508	.5008	6.633	1.997	1.134	2.265	4
5	4.484	.2230	.1005	.4505	9.954	2.220	1.422	3.157	5
6	6.053	.1652	.0693	.4193	14.438	2.385	1.670	3.983	6
7	8.172	.1224	.0488	.3988	20.492	2.508	1.881	4.717	7
8	11.032	.0906	.0349	.3849	28.664	2.598	2.060	5.352	8
9	14.894	.0671	.0252	.3752	39.696	2.665	2.209	5.889	9
10	20.107	.0497	.0183	.3683	54.590	2.715	2.334	6.336	10
11	27.144	.0368	.0134	.3634	74.697	2.752	2.436	6.705	11
12	36.644	.0273	.00982	.3598	101.841	2.779	2.520	7.005	12
13	49.470	.0202	.00722	.3572	138.485	2.799	2.589	7.247	13
14	66.784	.0150	.00532	.3553	187.954	2.814	2.644	7.442	14
15	90.158	.0111	.00393	.3539	254.739	2.825	2.689	7.597	15
16	121.714	.00822	.00290	.3529	344.897	2.834	2.725	7.721	16
17	164.314	.00609	.00214	.3521	466.611	2.840	2.753	7.818	17
18	221.824	.00451	.00158	.3516	630.925	2.844	2.776	7.895	18
19	299.462	.00334	.00117	.3512	852.748	2.848	2.793	7.955	19
20	404.274	.00247	.00087	.3509	1 152.2	2.850	2.808	8.002	20
21	545.769	.00183	.00064	.3506	1 556.5	2.852	2.819	8.038	21
22	736.789	.00136	.00048	.3505	2 102.3	2.853	2.827	8.067	22
23	994.665	.00101	.00035	.3504	2 839.0	2.854	2.834	8.089	23
24	1 342.8	.00074	.00026	.3503	3 833.7	2.855	2.839	8.106	24
25	1 812.8	.00055	.00019	.3502	5 176.5	2.856	2.843	8.119	25
26	2 447.2	.00041	.00014	.3501	6 989.3	2.856	2.847	8.130	26
27	3 303.8	.00030	.00011	.3501	9 436.5	2.856	2.849	8.137	27
28	4 460.1	.00022	.00008	.3501	12 740.3	2.857	2.851	8.143	28
29	6 021.1	.00017	.00006	.3501	17 200.4	2.857	2.852	8.148	29
30	8 128.5	.00012	.00004	.3500	23 221.6	2.857	2.853	8.152	30
31	10 973.5	.00009	.00003	.3500	31 350.1	2.857	2.854	8.154	31
32	14 814.3	.00007	.00002	.3500	42 323.7	2.857	2.855	8.157	32
33	19 999.3	.00005	.00002	.3500	57 137.9	2.857	2.855	8.158	33
34	26 999.0	.00004	.00001	.3500	77 137.2	2.857	2.856	8.159	34
35	36 448.7	.00003	.00001	.3500	104 136.3	2.857	2.856	8.160	35

40%   Compound Interest Factors   40%

	Single Payment		Uniform Payment Series				Arithmetic Gradient		
	Compound Amount Factor	Present Worth Factor	Sinking Fund Factor	Capital Recovery Factor	Compound Amount Factor	Present Worth Factor	Gradient Uniform Series	Gradient Present Worth	
	Find F Given P	Find P Given F	Find A Given F	Find A Given P	Find F Given A	Find P Given A	Find A Given G	Find P Given G	
n	F/P	P/F	A/F	A/P	F/A	P/A	A/G	P/G	n
1	1.400	.7143	1.0000	1.4000	1.000	0.714	0	0	1
2	1.960	.5102	.4167	.8167	2.400	1.224	0.417	0.510	2
3	2.744	.3644	.2294	.6294	4.360	1.589	0.780	1.239	3
4	3.842	.2603	.1408	.5408	7.104	1.849	1.092	2.020	4
5	5.378	.1859	.0914	.4914	10.946	2.035	1.358	2.764	5
6	7.530	.1328	.0613	.4613	16.324	2.168	1.581	3.428	6
7	10.541	.0949	.0419	.4419	23.853	2.263	1.766	3.997	7
8	14.758	.0678	.0291	.4291	34.395	2.331	1.919	4.471	8
9	20.661	.0484	.0203	.4203	49.153	2.379	2.042	4.858	9
10	28.925	.0346	.0143	.4143	69.814	2.414	2.142	5.170	10
11	40.496	.0247	.0101	.4101	98.739	2.438	2.221	5.417	11
12	56.694	.0176	.00718	.4072	139.235	2.456	2.285	5.611	12
13	79.371	.0126	.00510	.4051	195.929	2.469	2.334	5.762	13
14	111.120	.00900	.00363	.4036	275.300	2.478	2.373	5.879	14
15	155.568	.00643	.00259	.4026	386.420	2.484	2.403	5.969	15
16	217.795	.00459	.00185	.4018	541.988	2.489	2.426	6.038	16
17	304.913	.00328	.00132	.4013	759.783	2.492	2.444	6.090	17
18	426.879	.00234	.00094	.4009	1 064.7	2.494	2.458	6.130	18
19	597.630	.00167	.00067	.4007	1 419.6	2.496	2.468	6.160	19
20	836.682	.00120	.00048	.4005	2 089.2	2.497	2.476	6.183	20
21	1 171.4	.00085	.00034	.4003	2 925.9	2.498	2.482	6.200	21
22	1 639.9	.00061	.00024	.4002	4 097.2	2.498	2.487	6.213	22
23	2 295.9	.00044	.00017	.4002	5 737.1	2.499	2.490	6.222	23
24	3 214.2	.00031	.00012	.4001	8 033.0	2.499	2.493	6.229	24
25	4 499.9	.00022	.00009	.4001	11 247.2	2.499	2.494	6.235	25
26	6 299.8	.00016	.00006	.4001	15 747.1	2.500	2.496	6.239	26
27	8 819.8	.00011	.00005	.4000	22 046.9	2.500	2.497	6.242	27
28	12 347.7	.00008	.00003	.4000	30 866.7	2.500	2.498	6.244	28
29	17 286.7	.00006	.00002	.4000	43 214.3	2.500	2.498	6.245	29
30	24 201.4	.00004	.00002	.4000	60 501.0	2.500	2.499	6.247	30
31	33 882.0	.00003	.00001	.4000	84 702.5	2.500	2.499	6.248	31
32	47 434.8	.00002	.00001	.4000	118 584.4	2.500	2.499	6.248	32
33	66 408.7	.00002	.00001	.4000	166 019.2	2.500	2.500	6.249	33
34	92 972.1	.00001		.4000	232 427.9	2.500	2.500	6.249	34
35	130 161.0	.00001		.4000	325 400.0	2.500	2.500	6.249	35

45%

Compound Interest Factors

45%

	Single Payment		Uniform Payment Series				Arithmetic Gradient		
	Compound Amount Factor	Present Worth Factor	Sinking Fund Factor	Capital Recovery Factor	Compound Amount Factor	Present Worth Factor	Gradient Uniform Series	Gradient Present Worth	
	Find F Given P	Find P Given F	Find A Given F	Find A Given P	Find F Given A	Find P Given A	Find A Given G	Find P Given G	
n	F/P	P/F	A/F	A/P	F/A	P/A	A/G	P/G	n
1	1.450	.6897	1.0000	1.4500	1.000	0.690	0	0	1
2	2.103	.4756	.4082	.8582	2.450	1.165	0.408	0.476	2
3	3.049	.3280	.2197	.6697	4.553	1.493	0.758	1.132	3
4	4.421	.2262	.1316	.5816	7.601	1.720	1.053	1.810	4
5	6.410	.1560	.0832	.5332	12.022	1.876	1.298	2.434	5
6	9.294	.1076	.0543	.5043	18.431	1.983	1.499	2.972	6
7	13.476	.0742	.0361	.4861	27.725	2.057	1.661	3.418	7
8	19.541	.0512	.0243	.4743	41.202	2.109	1.791	3.776	8
9	28.334	.0353	.0165	.4665	60.743	2.144	1.893	4.058	9
10	41.085	.0243	.0112	.4612	89.077	2.168	1.973	4.277	10
11	59.573	.0168	.00768	.4577	130.162	2.185	2.034	4.445	11
12	86.381	.0116	.00527	.4553	189.735	2.196	2.082	4.572	12
13	125.252	.00798	.00362	.4536	276.115	2.204	2.118	4.668	13
14	181.615	.00551	.00249	.4525	401.367	2.210	2.145	4.740	14
15	263.342	.00380	.00172	.4517	582.982	2.214	2.165	4.793	15
16	381.846	.00262	.00118	.4512	846.325	2.216	2.180	4.832	16
17	553.677	.00181	.00081	.4508	1 228.2	2.218	2.191	4.861	17
18	802.831	.00125	.00056	.4506	1 781.8	2.219	2.200	4.882	18
19	1 164.1	.00086	.00039	.4504	2 584.7	2.220	2.206	4.898	19
20	1 688.0	.00059	.00027	.4503	3 748.8	2.221	2.210	4.909	20
21	2 447.5	.00041	.00018	.4502	5 436.7	2.221	2.214	4.917	21
22	3 548.9	.00028	.00013	.4501	7 884.3	2.222	2.216	4.923	22
23	5 145.9	.00019	.00009	.4501	11 433.2	2.222	2.218	4.927	23
24	7 461.6	.00013	.00006	.4501	16 579.1	2.222	2.219	4.930	24
25	10 819.3	.00009	.00004	.4500	24 040.7	2.222	2.220	4.933	25
26	15 688.0	.00006	.00003	.4500	34 860.1	2.222	2.221	4.934	26
27	22 747.7	.00004	.00002	.4500	50 548.1	2.222	2.221	4.935	27
28	32 984.1	.00003	.00001	.4500	73 295.8	2.222	2.221	4.936	28
29	47 826.9	.00002	.00001	.4500	106 279.9	2.222	2.222	4.937	29
30	69 349.1	.00001	.00001	.4500	154 106.8	2.222	2.222	4.937	30
31	100 556.1	.00001		.4500	223 455.9	2.222	2.222	4.938	31
32	145 806.4	.00001		.4500	324 012.0	2.222	2.222	4.938	32
33	211 419.3			.4500	469 818.5	2.222	2.222	4.938	33
34	306 558.0			.4500	681 237.8	2.222	2.222	4.938	34
35	444 509.2			.4500	987 795.9	2.222	2.222	4.938	35

50%

Compound Interest Factors

50%

	Single Payment		Uniform Payment Series				Arithmetic Gradient		
	Compound Amount Factor	Present Worth Factor	Sinking Fund Factor	Capital Recovery Factor	Compound Amount Factor	Present Worth Factor	Gradient Uniform Series	Gradient Present Worth	
	Find F Given P F/P	Find P Given F P/F	Find A Given F A/F	Find A Given P A/P	Find F Given A F/A	Find P Given A P/A	Find A Given G A/G	Find P Given G P/G	
n									n
1	1.500	.6667	1.0000	1.5000	1.000	0.667	0	0	1
2	2.250	.4444	.4000	.9000	2.500	1.111	0.400	0.444	2
3	3.375	.2963	.2105	.7105	4.750	1.407	0.737	1.037	3
4	5.063	.1975	.1231	.6231	8.125	1.605	1.015	1.630	4
5	7.594	.1317	.0758	.5758	13.188	1.737	1.242	2.156	5
6	11.391	.0878	.0481	.5481	20.781	1.824	1.423	2.595	6
7	17.086	.0585	.0311	.5311	32.172	1.883	1.565	2.947	7
8	25.629	.0390	.0203	.5203	49.258	1.922	1.675	3.220	8
9	38.443	.0260	.0134	.5134	74.887	1.948	1.760	3.428	9
10	57.665	.0173	.00882	.5088	113.330	1.965	1.824	3.584	10
11	86.498	.0116	.00585	.5058	170.995	1.977	1.871	3.699	11
12	129.746	.00771	.00388	.5039	257.493	1.985	1.907	3.784	12
13	194.620	.00514	.00258	.5026	387.239	1.990	1.933	3.846	13
14	291.929	.00343	.00172	.5017	581.859	1.993	1.952	3.890	14
15	437.894	.00228	.00114	.5011	873.788	1.995	1.966	3.922	15
16	656.814	.00152	.00076	.5008	1 311.7	1.997	1.976	3.945	16
17	985.261	.00101	.00051	.5005	1 968.5	1.998	1.983	3.961	17
18	1 477.9	.00068	.00034	.5003	2 953.8	1.999	1.988	3.973	18
19	2 216.8	.00045	.00023	.5002	4 431.7	1.999	1.991	3.981	19
20	3 325.3	.00030	.00015	.5002	6 648.5	1.999	1.994	3.987	20
21	4 987.9	.00020	.00010	.5001	9 973.8	2.000	1.996	3.991	21
22	7 481.8	.00013	.00007	.5001	14 961.7	2.000	1.997	3.994	22
23	11 222.7	.00009	.00004	.5000	22 443.5	2.000	1.998	3.996	23
24	16 834.1	.00006	.00003	.5000	33 666.2	2.000	1.999	3.997	24
25	25 251.2	.00004	.00002	.5000	50 500.3	2.000	1.999	3.998	25
26	37 876.8	.00003	.00001	.5000	75 751.5	2.000	1.999	3.999	26
27	56 815.1	.00002	.00001	.5000	113 628.3	2.000	2.000	3.999	27
28	85 222.7	.00001	.00001	.5000	170 443.4	2.000	2.000	3.999	28
29	127 834.0	.00001		.5000	255 666.1	2.000	2.000	4.000	29
30	191 751.1	.00001		.5000	383 500.1	2.000	2.000	4.000	30
31	287 626.6			.5000	575 251.2	2.000	2.000	4.000	31
32	431 439.9			.5000	862 877.8	2.000	2.000	4.000	32

60% Compound Interest Factors 60%

	Single Payment		Uniform Payment Series				Arithmetic Gradient		
	Compound Amount Factor	Present Worth Factor	Sinking Fund Factor	Capital Recovery Factor	Compound Amount Factor	Present Worth Factor	Gradient Uniform Series	Gradient Present Worth	
	Find F Given P	Find P Given F	Find A Given F	Find A Given P	Find F Given A	Find P Given A	Find A Given G	Find P Given G	
n	F/P	P/F	A/F	A/P	F/A	P/A	A/G	P/G	n
1	1.600	.6250	1.0000	1.6000	1.000	0.625	0	0	1
2	2.560	.3906	.3846	.9846	2.600	1.016	0.385	0.391	2
3	4.096	.2441	.1938	.7938	5.160	1.260	0.698	0.879	3
4	6.554	.1526	.1080	.7080	9.256	1.412	0.946	1.337	4
5	10.486	.0954	.0633	.6633	15.810	1.508	1.140	1.718	5
6	16.777	.0596	.0380	.6380	26.295	1.567	1.286	2.016	6
7	26.844	.0373	.0232	.6232	43.073	1.605	1.396	2.240	7
8	42.950	.0233	.0143	.6143	69.916	1.628	1.476	2.403	8
9	68.719	.0146	.00886	.6089	112.866	1.642	1.534	2.519	9
10	109.951	.00909	.00551	.6055	181.585	1.652	1.575	2.601	10
11	175.922	.00568	.00343	.6034	291.536	1.657	1.604	2.658	11
12	281.475	.00355	.00214	.6021	467.458	1.661	1.624	2.697	12
13	450.360	.00222	.00134	.6013	748.933	1.663	1.638	2.724	13
14	720.576	.00139	.00083	.6008	1 199.3	1.664	1.647	2.742	14
15	1 152.9	.00087	.00052	.6005	1 919.9	1.665	1.654	2.754	15
16	1 844.7	.00054	.00033	.6003	3 072.8	1.666	1.658	2.762	16
17	2 951.5	.00034	.00020	.6002	4 917.5	1.666	1.661	2.767	17
18	4 722.4	.00021	.00013	.6001	7 868.9	1.666	1.663	2.771	18
19	7 555.8	.00013	.00008	.6011	12 591.3	1.666	1.664	2.773	19
20	12 089.3	.00008	.00005	.6000	20 147.1	1.667	1.665	2.775	20
21	19 342.8	.00005	.00003	.6000	32 236.3	1.667	1.666	2.776	21
22	30 948.5	.00003	.00002	.6000	51 579.2	1.667	1.666	2.777	22
23	49 517.6	.00002	.00001	.6000	82 527.6	1.667	1.666	2.777	23
24	79 228.1	.00001	.00001	.6000	132 045.2	1.667	1.666	2.777	24
25	126 765.0	.00001		.6000	211 273.4	1.667	1.666	2.777	25
26	202 824.0			.6000	338 038.4	1.667	1.667	2.778	26
27	324 518.4			.6000	540 862.4	1.667	1.667	2.778	27
28	519 229.5			.6000	865 380.9	1.667	1.667	2.778	28